A MAKESHIFT PLANET

By: Andrew Fischer-Gledhill

COPYRIGHT

Published by

BookFuel™

Table of Contents

1
MAKESHIFT

The first test you fail is the one you never take off the paper and into life.

In the daylight, beneath a patch of shifting shade, a boy of about fourteen sat twisting blades of grass into a spiral between his fingers. He studied the shape as it formed, as if it held some quiet meaning only he could understand. A breeze rushed past with a soft *whoooosh*, lifting his hair toward the sunlight and drawing his eyes closed for a brief moment.

Behind him stood a house that didn't quite belong.

It sat on the outskirts of a vast city pressed against the mountains, where the skyline in the distance rose like a steel forest. Somewhere beyond that horizon lay one of the most industrious districts in the region—a place driven by machines, factories, and constant motion. Yet here, between that relentless world and the quiet stretch of open land, there was still a pocket of calm.

The boy was an Arlenish secondary school student. And a poor one at that.

He sat cross-legged in the grass, waiting for his grandfather to return home.

His grandfather, now in his late sixties, worked at Local Tech, a robotic repair plant not far from the city. Inside, everything was white, the walls, the floors, even the uniforms worn by the technicians. They moved through the facility in full-body suits and matching masks, resembling astronauts preparing for a mission.

The design wasn't for appearance—it served a purpose. In such a spotless environment, even the smallest flaw or trace of contamination became immediately visible. Every machine that entered the facility had to be handled with care, as if it were something irreplaceable.

Despite the precision and importance of the work, the wages were meager—sometimes painfully so. Still, it was one of the few jobs available in the region, and one the old man could still manage.

More importantly, it was work that had not yet been taken over by automation.

For now, at least, it was still his.

On his days off from school, the boy helped however he could. He traveled into the nearby metropolis to buy, trade, and bargain for old items—antiquing, as his grandfather called it. It wasn't much, but sometimes it brought in just enough to help them get by.

He often found himself thinking about the city.

If it was like a tiered diamond—shaped by centuries of pressure, pulling people inward to make it shine—then what, he wondered, became of the outskirts?

The boy himself was of average build, perhaps slightly on the slender side. Once pale, his skin had begun to tan from spending long hours outdoors. He stood just over five feet tall and occasionally experimented with different hairstyles, though for now his hair remained in its natural shade somewhere between light brown and blond.

At fourteen, he was often told he was ahead of many of his peers. Still, he was rarely taller than them—and if he was, it was never by much.

He and his grandfather had moved frequently over the years before finally settling in a small suburban town that

bordered on rural life. The town existed largely in support of the nearby city, populated by contractors and workers whose livelihoods depended on it.

The houses were spaced comfortably, open enough to avoid feeling crowded, but not so distant as to resemble farmland. Beyond the town's edges, however, fields and farms stretched across the horizon.

Around their home, nothing was planted without purpose. The flowers and greenery were carefully chosen by his grandfather—some edible, some medicinal, and others cultivated for practical uses such as producing dyes for clothing.

Most of the neighborhood lawns were grown from engineered

grass seed, designed to reach a selected height and remain there indefinitely without the need for mowing. Some were select-a-height electronically regulated grass seed. Others took an even easier route, replacing grass entirely with synthetic turf.

Their home, however, stood apart.

The backyard appeared almost barren compared to the neatly maintained lawns nearby. Yet the old saying still holds true the grass is always greener on the other side. In their case, it quite literally was. The front yard, at least, was greener than most.

The neighbors themselves were just as varied. Each seemed to carry a distinct personality, though their homes rarely reflected it. Styler's house sat between two properties, one belonging to people you might grow to like, the other to people you might just as easily grow to dislike. Across the street lived another neighbor who, Styler sometimes joked with, looked as though they were simply waiting for the day they could dance on someone's grave.

On either side, the neighboring houses appeared slightly nicer from the outside. Yet something about them felt hollow, as if they stared outward without life. They lacked the sense of character and presence that filled Styler's home.

At sixty-seven, Styler's grandfather still worked. He did so with quiet reluctance, burdened more by necessity than desire. Though he complained from time to time, he endured it because he had to. And no matter how long or tiring the day had been, he always looked forward to coming home.

The boy had a name.

In the distance, an older man's voice rang out with familiar warmth.

"Styler!"

The Arlenish boy turned at the sound. His name was Styler—though more often than not, people misheard it as Skyler, Stiller, Stellar, or something close enough to be wrong. By now, he was used to correcting them.

A moment later, he heard the soft approach of a vehicle along the road. His grandfather had returned home in his 2116 Morrison Mod

5—once the finest model of its year, though noticeably outdated by the standards of 2137.

Styler liked the car more than he admitted. The name "Mod" reminded him of game modifications—small changes that made something standard feel personal. To him, the vehicle carried that same idea. It felt customized, almost like it belonged uniquely to them, even if he wasn't much of a gamer himself.

For its time, the Mod 5 was impressive. Its roof, made of rigid electro-synthetic glass, provided a completely unobstructed view in every direction. With the flip of a switch, an electrical current could render it opaque. Not that visibility

mattered much—like most vehicles of its era, it operated autonomously.

The turbine engine required minimal energy while producing remarkable torque and power. It also recharged wirelessly, drawing energy from transmitted signals sent through power towers connected to the same grid that supplied ordinary homes.

More than that, the vehicle could move in nearly any direction. With its hover capability, it wasn't limited to roads—it could glide forward, backward, sideways, or diagonally through the air with ease.

The vehicle slowed and came to a stop. Styler's grandfather stepped out, and the car quietly guided itself into the small one-car garage beside the house.

Inside, everything was in order.

The garage was kept with near-perfect precision. Every tool had its place, every surface was clear, and nothing was wasted. Harold often repeated his simple rule:

All in line. All in order. No waste.

Tucked within the garage was a compact tool room and workspace—one that Harold believed could one day become something more. A place not just for repairs, but for invention. Styler raised his voice toward him.

"Harold!"

Styler refused to call him Grandpa. He preferred to be treated like an adult, and using Harold's name made that feel more real.

Harold answered with a brief shout of acknowledgment, his voice deep and raspy.

Styler dropped the blades of grass he had been twisting between his fingers and waited as Harold made his way toward the house. The old man carried a small, worn leather bag embossed with his initials—nothing more.

As the wind picked up slightly, Styler followed him inside.

The makeshift house stood apart from the rest of the neighborhood. Some might have called it unkempt, even impoverished, but to Styler it had character—something the surrounding homes seemed to lack. Nearly every part of it had been modified by Harold. The changes were unusual, sometimes even strange, but each one had a purpose.

Harold believed a home should make life easier.

One of the house's most distinctive features was its engine. Built from salvaged parts of early 2000s fossil-fuel vehicles, it powered much of the home's internal systems. Turning the key beside the front door brought the mechanism to life with a low rumble, triggering the lights and regulating the temperature so the house warmed or cooled within moments.

Inside, the space felt lived in. Quilts hung within easy reach, suggesting the quiet presence of someone who had once cared for the home in a different way. Yet alongside that warmth were strange, improvised devices—homemade systems designed to simplify daily life.

Styler and Harold had built many of them together. Over time, Harold had taught him to repair, reuse, and invent whenever possible. They were hobbyist inventors at heart.

There had never been a father. No mother.

Only Harold.

Recycling was one of Harold's firm principles. As Styler stepped inside, he remembered it was his turn to take it out. They used clear bags so everything inside could be seen—both to keep things organized and to ensure nothing ended up where it shouldn't.

Most waste, however, never made it that far. In the kitchen, an automated disposal system sorted materials and sent them to a processing station for recycling.

The house produced much of its own energy. Its windows were made of translucent solar panels, capturing sunlight while still letting light pass through. Even the fence served a purpose, constructed from darker solar panels designed to absorb more heat and generate additional electricity.

Rainwater and melting snow were directed through a customized gutter system along the roof. As the water moved through it, small turbines spun, generating power. Even wind passing through the system contributed to the house's energy. As Harold liked to say, the house even consumed fog.

Styler moved ahead of his grandfather and reached the entryway first. He turned the key mounted beside the door.

A click.

Then a low vroooooom echoed through the structure as the house came alive.

Inside, the walls and ceiling were filled with a collection of curious inventions. Suspended overhead, a gas-based chemistry apparatus glowed with neon light as gases moved through a network of improvised containers. Some pieces were proper laboratory glassware, while others had been fashioned from everyday materials—a metal tin can, a plastic bag, even folded paper—each one carefully arranged and softly illuminated.

Along one wall stood a row of seats salvaged from an old aeronautical cabin. They had been reupholstered with textured, embossed fabric, with small mechanical elements subtly built into them. Nearby sat two additional chairs reclaimed from aerospace vessels, still fitted with the heavy harnesses once used to secure passengers during flight.

Styler moved past it all and headed straight for the kitchen.

He passed a round table equipped with built-in lighting and a heating element at its center, allowing meals to be prepared or warmed directly on its surface.

At the sink was one of the house's more unusual features.

By this point in time, more than nineteen types of utility lines existed—though not every home had access to all of them. Harold, however, had managed to install one of the rarer ones.

The sink had two pipes. One was for water.

The other was not. A food pipe.

Positioned beside the faucet—mise en place, as the French would say—it allowed food to be delivered directly into the home. With a simple activation, it could dispense items such as berries, soups, stews, or protein pellets.

Styler paused, looking at it. Hungry.

Food pipe.

It made perfect sense.

To maintain sanitation, the system was automatically flushed with lemon juice every other day. With a setup like this, there was little need for traditional grocery shopping.

Food could simply arrive at home.

Not every home or neighborhood had a food pipe installed, but Harold believed it was a worthwhile investment. Years earlier, when pollution from vehicle emissions had overtaken several major cities, many communities began adopting systems that reduced transportation and simplified food distribution. The food pipe was one of those solutions.

Even so, it had its limitations. Only certain types of food could pass through it—berries, protein pellets, soups, stews, and other simple meals designed for easy delivery.

Styler activated the dispenser and released a portion of protein pellets into a container. He filled a reusable cup with water. Normally, he would have poured himself a glass of his favorite drink—Zox, the color-changing beverage that shifted its flavor with every sip—but it was gone.

He sat at the table while Harold unpacked his work bag. As he ate, his thoughts drifted.

He was six years old again.

Back then, his mother was still alive. So was his grandmother. The house had been louder—filled with voices, movement, and laughter.

They had gathered around the table for a meal. Harold, as always, had been eating with remarkable focus. By the time Styler glanced up from his plate, most of the food had already disappeared.

His grandmother broke the silence.

"What are you doing?" The table went still.

Harold paused mid-bite, wiping his hands as if he had been wearing a bib, and looked up with practiced innocence.

"You're telling me you ate all of it?" she said. "All that food? What is this—you snooze, you lose?"

Styler's mother leaned in with a grin.

"Maybe it's more like the early bird gets the worm."

Harold bent slightly toward young Styler and whispered, his voice low and amused,

"That's a life lesson for both of us." The memory slipped away.

Styler's gaze shifted toward the yard outside. Near the door sat the unfinished remains of the latest project he and Harold had been working on—a vehicle they had been welding together piece by piece. It was still far from complete.

Inside, on a nearby table, miniature versions of the design floated in a shallow water basin. They were prototypes— used to test balance and movement. Styler watched them for a moment, imagining what the finished machine might one day become.

After a while, he tried to get Harold's attention. "H—"

"Rolddddddd!"

He stretched the name out deliberately, then added with a grin, "HarHarHar!"

Harold looked over, shaking his head with a faint smile.

"As if anyone could pronounce your name properly,

Styler."

A bark suddenly echoed from outside.

Harold's dog, Fire, had returned.

Fire was a genetically engineered husky with long fur

that shimmered in shifting shades of red. The patterns in his coat seemed almost alive, subtly changing as he moved. Despite the thickness of his coat, the breed had been designed to remain comfortable even in warmer climates.

He preferred being outdoors whenever possible, which suited Harold just fine. Fire shed constantly, and keeping most of that vivid red fur outside made life easier.

Moments later, the door burst open.

Fire trotted inside, muddy paws dragging dirt across the floor.

"Smelly Fire—go outside!" Harold shouted.

The glowing red dog paused, staring at him without reaction.

Fire, after all, didn't understand a single word of Arlenish.

With a quiet sigh, Harold stepped forward and guided him back outside.

The house itself had no television. What had once been the center of entertainment—television and streaming—had gradually faded into irrelevance. Over time, people had simply grown bored of it.

Instead, most relied on holocasters.

Holocasters were compact, multifunctional devices worn on the head, resting lightly like glasses or a monocle. Despite their size, they were capable of an astonishing range of functions.

Through a holocaster, a user could access information, communicate instantly, or connect to countless external systems—machines, networks, devices, even genetic databases. By focusing on a location, the device could retrieve detailed historical records of events that had occurred there.

More advanced systems could go even further.

Using somewhat supernatural technology, holocasters could recreate scenes from centuries past—moments that existed long before cameras or recorded data—rendering them as fully immersive visual and audio experiences.

Communication worked just as seamlessly. To contact someone, a user only needed to think of them. The device recognized mental imagery—faces, emotional impressions, even biological identifiers—and established the connection instantly.

There were also features that few people used anymore.

In rare cases, holocasters could allow communication across

time—connecting users with individuals from the past, or even the future, provided enough data about them existed.

For most, however, the device had simply replaced everything television once was.

Nearby, Styler stood at a worktable illuminated by a cluster of glowing vessels suspended overhead. Their soft light reflected across scattered tools and components.

He picked up a soldering iron and began working on a motherboard—just one small part of the larger vehicle he and Harold were building together.

As he worked, a thought lingered.

Each piece was small. Insignificant on its own. Yet every connection mattered. Without them, nothing would function.

In a way, it reminded him of people—each one a small part of something much larger.

Styler continued soldering the delicate lines, unsure if he was doing it entirely right. Mostly, he followed what Harold had shown him.

After a moment, he slipped on his holocaster to get a closer look. The display zoomed in, magnifying the circuitry so he could see his work more clearly.

Then a notification appeared. Styler frowned.

"Wow… these holocaster pop-ups are terrible," he muttered. "They're like hallucinations."

He dismissed the notification and refocused the display, continuing to solder the connections with careful concentration.

Before long, fatigue set in.

Styler leaned back into one of the reclaimed chairs—an old aeronautical seat Harold had restored years earlier. He powered down the holocaster and pulled a quilt his grandmother had once made over his shoulders.

Within moments, he drifted into sleep.

Moments later…

Styler lifted the holocaster from his head.

The device had been running in past-view mode.

What he had experienced was not a dream, but a recorded

reconstruction—a memory from earlier years, replayed through the holocaster's time-viewing function. The technology allowed users to observe moments from the past, the present, and, in rare cases, even projections of the future. As he removed the device, the sensation lingered—strangely similar to waking from deep sleep.

Morning light streamed into the room.

Styler blinked, stretched, and sat up slowly, letting the feeling pass as the day settled in around him.

Yet something remained. A quiet restlessness.

He found himself wanting more—something beyond what he had, beyond the routine of his days. The world around him felt vast, full of possibility, yet somehow distant.

The boy struggled to form connections with others—or at least the right ones. And without them, meaning felt just out of reach.

2

TIME FOR CLASS

If you stop to see them, the choices are always there—right in front of you. No one else can make them for you. Wherever you go, you carry yourself with you. Your thoughts, your past, your scars—they follow.

You are always you.

The treetops stretched overhead, their leaves a vivid green, matching the grass below. Life existed everywhere—abundant and continuous—while people, in comparison, felt few and fleeting within it.

In class, the students gathered for a headcount before heading outdoors on a clear, beautiful morning. They were preparing to visit the nature preserve just beyond the school grounds, a space reserved exclusively for the students and faculty of the secondary school Styler attended.

"One, Michael. Two, Michelle…"

Styler listened as the names were called, his thoughts drifting.

Great. Numbers before names.

"Three, Taylor. Four, Styler. Five, James. Six, Ariel. Seven, Jennifer. Eight, Dave. Nine, Robert. Ten, Sean. Eleven, Jessica."

The teacher finished the count. All eleven students were present.

As the teacher turned his back, Styler's gaze shifted to a larger boy—Sean.

Sean noticed.

A smirk crept across his face.

"Hey, number four," he called out, glancing at Styler's equipment. "You look like you ate too much."

The remark was mocking—an obvious jab, considering Styler's thin frame.

He continued.

"What are you, foster home? Or homeless shelter?"

A few nearby students reacted, some amused, others uncomfortable. Styler said nothing. He bit down lightly on his lip, holding his composure.

The words landed closer than Sean probably realized.

Poverty was more widespread now than many admitted. Even with automation, the growing population had only intensified competition, leaving many behind.

Sean stepped closer, his attention shifting to a stain on Styler's light blue hoodie.

"What—did you find that on the side of the road?" He didn't stop.

"Hey. I'm worth more than you. You hear me?"

Around them, students held their holocasters—small personal devices powered by atomic batteries. After years of research, these batteries had replaced traditional power sources. They didn't need charging or replacement. Each one, no larger than the tip of a nail, functioned like a miniature reactor—scaled down through advanced engineering.

Nearly every device in use depended on them.

Before Styler could react, Sean reached forward, snatched his holocaster, and threw it to the ground.

It struck hard.

The device cracked. A sharp spark flickered—then another—before it began to rupture.

Dave flinched. "What if he can't afford another one?!"

Styler stared at Sean, his expression tightening—anger held just beneath the surface.

"There's nothing I've ever done to you," he said quietly.

Sean laughed.

"Being poor is enough," he replied. "Go stink somewhere else."

With that, he turned and walked away.

Around them, the rest of the class shifted their attention back to the teacher, as if nothing had happened.

It was the year 2137—spoken simply as two one three seven. The word year had long since become unnecessary.

The world had changed.

New cultures had emerged rapidly, shaped not over centuries, but within decades—accelerated by technology. Language had evolved alongside them. The global standard was Arlenish, spoken, written, and recorded almost everywhere. It had become so widespread that most people simply called it Common. Even this book, in its original form, had been written in Arlenish before being translated.

Still, not everyone spoke so openly. There was another language—Nadjin.

It existed in fragments at first, without a true population. But over time, a small, deliberate community formed around it. Its speakers guarded it carefully, sharing it only with a chosen few.

Where nothing exists, something eventually forms.

Where something feels inauthentic, something real often follows.

Or at least, that's what people believed.

Robert, one of Styler's classmates, knew only a single word of Nadjin. Even so, he turned toward Sean and muttered it under his breath:

"Jecha."

Sean didn't react. He didn't understand it.

That, perhaps, was the point.

The teacher gathered the class and began leading them back inside. Styler followed, pulling his light blue hoodie tighter around himself. It was the most practical thing he owned—affordable, durable, and warm.

Above them, commercial aircraft passed through the daylight sky.

They were far from ordinary planes.

Powered by compact reactors, these aircraft could travel not only through the air but also into space—and even underwater. Their wings could retract seamlessly into the fuselage, allowing for vertical takeoff and landing. With nearly limitless energy, they rarely needed to land at all unless necessary.

They were one of many advancements born from the energy era.

Back in the classroom, Styler moved quickly.

He headed straight for a multifunctional fabrication unit—a machine capable of producing objects across multiple dimensions, from simple printed components to fully integrated devices.

Within moments, he began assembling a new holocaster.

"It's a little trick I've learned," he said to a nearby classmate, focused as he worked.

Then, with a faint grin, he added,

"So I'm number four. That's not bad… at least."

He fitted the newly made holocaster onto his head, adjusting it carefully.

"I hope this works."

A moment later, his expression brightened.

"Couldn't even buy better."

From across the room, the teacher had been watching.

He stepped closer, clearly impressed.

"What do you want to be when you grow up?" he asked.

Styler didn't hesitate.

"Wealthier than I am now," he said plainly. Then, after a brief pause, "I'd like to be able to afford anything. I guess… an entrepreneur."

The teacher nodded, understanding more than Styler had said.

"I know things aren't easy for you," he replied. "If you ever need guidance, I'm here."

Styler took his seat beside a girl named Ariel. Her eyes were green—and so was her hair. It wasn't unusual anymore.

Genetic variation had entered a new phase. Traits that once seemed rare had become more common, even dominant. Green eyes had existed long before, but now they often appeared alongside naturally green hair—a pairing that seemed to have emerged from those same genetic markers.

Brown eyes and hair still existed.

But they were no longer the only standard.

There was a popular fable about how green hair first

appeared—that an unknown couple, conceiving at dusk beneath a green-tinted moon, gave rise to the new trait. Whether anyone believed it or not, the story persisted.

In reality, the change had been far more deliberate.

About a quarter century earlier, a group of scientists had discovered how to manipulate the genes responsible for eye and hair color. Green eyes, which had existed for thousands of years, were used as a foundation to develop new variations—most notably a deep green hair that could shift in tone depending on environmental conditions. The effect, however, remained rare.

Other variations followed. Scientists successfully paired red hair with red eyes, further expanding the visible range of human genetics.

Styler arrived home early, frustration written across his face.

The words Sean had thrown at him lingered, echoing louder now that he was alone. He carried them with him, turning them inward, letting them settle deeper than they should have.

He pushed through the door with more force than necessary.

Outside, Harold was tending to the front yard. He noticed immediately.

Without saying a word at first, Harold picked up two sticks and clapped them together sharply—like the old film clapboards used to mark a scene.

"Take two," he called out.

It was his way of telling Styler to reset. To pause. To try again.

An old habit, maybe—but not a meaningless one. After a moment, Harold stepped closer.

"I'm heading out for my evening shift," he said. "You going to be alright while I'm gone?"

Styler didn't look at him. "I'll be me," he murmured.

Harold nodded, accepting the answer for what it was.

"That's alright, then."

He climbed into the Morrison Mod 5, lifting it smoothly off the ground before guiding it toward work—toward the city, though not quite within it.

Styler remained behind.

For a moment, he stood still, unsure of what to do next. Eventually, he moved to the kitchen and pulled a portion of protein from the food pipe. He ate without much thought, his mind still caught in the weight of the day.

He wanted things to feel different. Better.

They didn't.

A dull discomfort crept in. He felt cold, then warm, then tired—like the beginnings of a fever. Absentmindedly, he rubbed at an old scar, as if grounding himself.

After a moment, he put on his holocaster and switched to its medical interface. A menu appeared before him.

He selected: fever and drowsiness.

The device responded immediately, emitting a low, calming tone while directing a subtle stream of energy through his ear. The sensation was gentle but precise. Gradually, the discomfort faded. His body steadied. His mind cleared.

Feeling more alert, Styler turned his attention back to the model he and Harold had been working on—the small prototype of the vehicle waiting unfinished outside.

Harold, meanwhile, was already on his way.

He traveled along the triway—a long, three-lane highway constructed from brown-toned synthetic materials that resembled natural earth. The design was intentional, meant to blend with the environment and reduce disruption to surrounding wildlife.

It was the route he took every day.

Along the way, he passed various automated systems at work. Among them were units known as Farsers—coordinated robotic machines designed to manage vegetation along the roadway.

They moved with precision, clearing trees and overgrowth while simultaneously analyzing the material they cut. Using vast databases, they could predict the internal structure of wood—its grain, its knots, even its potential uses—before making a single cut.

Some materials were processed into construction lumber. Others were designated for furniture, composites, or fuel.

The systems could even estimate the scent of the wood before it was cut.

Similar technologies had been adapted for geological use, including systems like the AI Geological Excavator (AGE), capable of predicting mineral structures beneath the surface before excavation began.

Harold barely noticed them anymore.

His focus remained ahead—on the distant outline of the city.

He always knew his exit by sight alone, when the massive skyline began to rise along the horizon.

Over time, humanity had reshaped the planet in ways few had once imagined. The construction of cities, the movement of mass, and the sheer scale of human activity had subtly altered the Earth itself—affecting its balance, even slowing its rotation.

Now, most of that work was handled by machines.

From space, the planet looked extraordinary—its surface illuminated by endless networks of cities stretching across continents.

And yet, the question remained. What had it cost?

The planet had been reshaped—disfigured, defaced, and deformed in the process of building cities. What remained was something unfamiliar, almost unrecognizable. A world pushed so far from its natural state that it felt… out of place.

A planet others might not want to be near.

There were stretches where pollution and waste replaced everything else—regions where nature had vanished for thousands of miles. What had once been balance was now excess.

What had been done?

And what would be left in the years to come?

It almost felt as though the planet itself might one day try to realign—to return, somehow, to harmony with others.

But perhaps it would no longer be welcome.

Harold arrived at Local Tech, the robotic repair plant where he worked.

The facility stood two stories tall just off the triway, within sight of the rail lines that led into the city. Outside, piles of discarded parts and retired machines formed uneven heaps—remnants of systems that had once been essential.

The building itself was stark white, its exterior designed to reflect light even in low visibility. Above the entrance, a bold sign stretched across the front:

"Build up locally with Local Tech!"

Harold moved quickly beneath it, entering through the employee access and heading inside to change into his uniform before taking his place at his usual station.

He knew robotics inside and out.

And yet, he had never been formally educated.

Harold had never believed much in traditional schooling or university systems. Even so, Local Tech had taken a chance on him—a rare decision. His ability spoke for itself. He could perform complex calculations in his head and was fluent in more programming languages than most formally trained engineers.

At his console, he began working on an excavation unit.

Then something went wrong.

The machine jolted, its bucket arm spinning erratically

before striking another unit and knocking it off balance. The foreman's voice cut sharply across the floor. "Call a PURR!"

PURR—short for Personnel Unit Robotics Repairman. The nickname came from the idea that a good repairman could get a machine running smoothly again—purring, like a kitten.

Harold didn't hesitate. "At the ready!"

He began entering commands into his console, linking wirelessly to the malfunctioning unit. Line by line, he worked through the system, isolating the error and overriding the movement sequence.

Within minutes, the machine stilled.

A coworker nearby called out, half-joking,

"Harold, when are you going to retire and give the rest of us a chance?"

Harold didn't look up.

"Keep asking me that," he replied, "and it's going to be a long shift."

The coworker laughed.

"Long enough for you to retire into it."

Back at home, Styler lay on the floor with Fire.

The dog's weight rested across his chest as Styler absentmindedly held out treats, letting Fire take them one by one. The rhythmic motion was calming, but his thoughts were elsewhere.

He couldn't shake the feeling.

The quiet fear of losing Harold.

It lingered beneath everything—unspoken, but constant.

Styler stared upward, feeling hollow, as though something inside him had been emptied out. Alone in the house, the silence felt heavier than usual.

At the plant, Harold finished recalibrating the excavation unit. He restored it to its default settings, then made a few subtle adjustments—improving its accuracy, increasing its trenching efficiency.

He took pride in that.

No job had ever felt beneath him. Over the years, he had worked in places and conditions most people wouldn't choose—sorting systems, repair lines, tasks others avoided. Still, there was always a sense that he was capable of more.

Much more.

His ancestry traced back to the early foundations of the city itself. By that measure, he could have held a position of influence—a desk job, a public role, something respected.

Instead, he was here. Working.

Fixing.

Enduring.

The foreman's voice rang out once more.

"Closing out!"

The signal that the facility was shutting down for the night.

The lights throughout the plant began to dim in

sequence, the hum of machinery fading with them. Like clockwork, the system wound down—ready to begin again in the morning.

Harold's shift was nearly over. Before leaving, he stopped at one of the facility's vending units and picked up dinner to take home.

On the drive back, he pushed the vehicle a little too fast while using the assisted driving feature. A flashing red warning appeared across his heads-up display—an automated speeding ticket. Another alert followed, warning him of an illegal lane shift.

Harold exhaled sharply and tapped the steering wheel.

"Oh, it's not like the old days, is it?" he muttered. "Oh, thank you."

At home, Styler was still awake.

He sat with his holocaster on, scrolling through his homework program. Not working—just looking.

Surveying, he called it.

The door opened a while later as Harold returned, a little lighter in the pocket thanks to the ticket. He guided the dark-colored Mod 5 into the garage and stepped inside.

"Styler, did you eat?" Harold called out. Styler glanced up with a faint smirk. "Why? What've you got for me?"

Harold set the packaged meal down on the round kitchen table.

"Easy Vend Dinner—the finest," Styler said dryly. "You know, I could live off condiments."

"This is a little more nutritious," Harold replied.

They sat together and unwrapped the meal.

Murton fruit macaroni and cheese, served with steak and broccoli.

Murton fruit—a dry, bitter fruit imported from the second quadrant—was commonly used as a pasta substitute. By this point in time, importing food from other planets had become routine.

Even with centuries of culinary advancement, though, few things had surpassed macaroni and cheese.

The vending system at Local Tech was no ordinary machine. It produced meals of near gourmet quality, based on recipes developed and endorsed by some of the most respected chefs of the era—including Louis Riley, a culinary figure known for traveling across systems in pursuit of the perfect bite.

Styler reached for the water pipe, filling two measured

containers—one for himself, one for Harold.

Measuring had become second nature to people. Water intake, food portions—everything was tracked. It was believed to sharpen discipline and reinforce scientific thinking, much like the routines once followed by astronauts.

In this time, everyone was treated as a potential astronaut.

From a young age, students were required to meet physical and health standards, including annual Astronaut Fitness Tests. Styler had struggled with those more than once—just barely meeting the requirements needed for future clearance.

Harold raised his glass slightly.

"To a hard-earned meal."

"Cheers," Styler replied. "Cheers."

They ate in relative quiet.

After a few minutes, Styler leaned back, full and a little drowsy. He hesitated, then passed his holocaster over.

"Can you help me with this?" Harold glanced at the screen.

"Biology?" he said. "I'm basically a mechanic." "Well… it's kind of mechanics," Styler replied. "Just…

of the body. And plants. And everything else."

Harold gave a small, apologetic shake of his head.

"You've got a digital book for it, right? You're going to have to read through it. I want to help—but there are things I just don't know."

Styler nodded, accepting it. Then, after a pause—

"How was the drive home?"

Harold hesitated for a fraction of a second.

"Drove a little fast," he said. "But it was fine." Styler smirked.

"Careful. You might get a sataticket." Harold gave a short nod.

"Yep."

3

IN A DAY'S WORK

The day was overcast. A storm was building.

The sky, once bright blue, had begun to darken. The same sky that mirrored the ocean. The same sun that once cast a warm yellow over the sand now hid behind thickening clouds. A low rumble echoed in the distance, and flashes of lightning cut across the sky—brief, violent, alive.

Styler listened to the weather broadcast through his holocaster as he traveled.

"Oh, it's a stormy one out there—especially in the east, the Query region. If you're out there, you'll need an extra-large umbrella. In the city, expect heavy cloud cover. Outskirts to the west will see brief rain, while the east shares cloud patterns with the city. Let's make it a great one! Sandra of the North East—now over to Sandra for the daily sessions while I take a nap."

By the time the broadcast ended, Styler had already entered the city.

He had taken the rail alone.

The track carried him across a sprawling bridge that stretched over the ocean. It wasn't an ordinary structure. Instead of traditional supports, skyscrapers rose directly from the water, anchoring the bridge in place. Entire residential towers lined both sides, with restaurants at their base—places where people could pause mid-journey to eat above the sea.

As the train moved deeper into the city, clouds thickened overhead until the skyline blurred from view.

Styler adjusted his slim-fit Suto backpack. Inside, carefully packed, were pieces of aluminized jewelry from the early 2010s—mass-produced at the time, but now rare enough to be valuable. He planned to pawn them, as he often did.

Once off the rail, he made his way toward his usual pawn shop.

Above him, a satellite projection cut through the clouds, forming a massive advertisement in the sky. Semi-translucent shapes shimmered into place, glowing like floating holograms.

"AI gun turrets!!! AI gun turrets!!!"

The cloud shifted, forming an animated display—a security turret firing at simulated vehicles attempting to breach a perimeter.

"Because it's been proven—be afraid of guns that move!!!"

The imagery changed again. A turret aimed at a front door, firing at a staged intruder who collapsed dramatically.

"AI gun turrets!!! Because guns that don't move are okay for you!"

The message lingered overhead, loud and intrusive. Styler kept walking.

The pawn shop he frequented sat in a part of the city most people avoided—crowded, tense, and unpredictable. It wasn't a place to stay longer than necessary.

He moved along the street beneath a network of dark gray catwalks suspended above the ground. The structures were assembled in interlocking rectangular sections, fitting together with precision—almost like pieces of a toy set. Their beveled edges locked tightly, forming a continuous protective grid over pedestrian pathways.

Across the street, workers in white compression uniforms were assembling another section. The fabric was deliberately bright—any grease, dust, or stain would be immediately visible, leaving no room for carelessness.

They worked in coordination, using retractable pulley systems to position each piece. The components hovered briefly in place before being guided into alignment.

Curiously, no CatwalkBots were in use.

Instead, the workers relied on manual coordination—though the materials themselves were transported on automated hover carriers, drifting silently between them.

Styler watched for a moment as he walked. Then continued on.

High above the city, construction never stopped.

Robotic units moved in coordinated patterns along the rising frame of a new skyscraper—climbing vertically and shifting horizontally with precise control. Each machine carried structural components, lifting metal beams and hauling panes of synthetic glass into position. With careful calibration, they placed each panel exactly where it needed to be.

These were 2131 Attec Climb-and-Carry units—designed to operate at heights no human could safely reach. At those elevations, oxygen levels dropped to dangerous limits. For a person, one mistake could mean their last breath.

For the robots, it meant nothing.

They worked without pause, hauling materials upward, assembling the building piece by piece. Human workers remained on-site, but mostly as supervisors—accessories to the machines doing the real work.

Buildings now stretched for miles—both into the sky and deep beneath the ground.

What most people didn't realize was what lay below.

Buried within the foundations of some of these structures were aligned fusion rocket systems—carefully embedded, hidden from view. They were part of long-term plans to make entire buildings mobile. One day, even the largest airtight skyscrapers could be launched and transported to other planets.

The city wasn't just expanding.

It was preparing to leave.

Space launch capable skyscrapers.

Elsewhere, planners had begun repurposing older parts of the city. In poorer districts, where aging skyscrapers stood unused, rooftops were being connected—linked together into massive platforms.

Runways. Landing zones.

An airport in the sky.

The concept had emerged after the third space race, and locals had given it a name:

The Skyport. Rain began to fall.

The shift in weather triggered an automatic response. The construction units paused, recalculating environmental conditions.

One by one, they retreated in sequence, moving into designated shelter zones to avoid damage from prolonged exposure.

Even machines knew when to step back.

Styler arrived at his usual pawn shop and stepped inside.

The interior was familiar. On the front counter sat a small perpetual motion device, spinning endlessly and producing a faint, steady hum.

Behind the counter, the owner looked up immediately.

"Styler!" he greeted with a wide smile. "What've you got for me today?"

Styler slipped off his slim gray backpack.

"You're going to like these, Jake."

He reached inside and carefully pulled out the items—aluminized jewelry from the 2010s. Two necklaces. One bracelet.

Jake's eyes lit up.

"Oh, these are nice," he said, examining them. "You don't see these anymore. Hard to find. Where'd you get them?"

"An older lady in my neighborhood," Styler replied.

He didn't mention the yard sale—or how little he had paid for them.

Jake nodded, turning the pieces in his hands.

"So," he said, glancing up, "how many hour credits are you looking for?"

Currency had changed over time. After years of inflation and financial instability, traditional systems had collapsed. In their place came hour credits—a currency based on units of work.

Minutes. Hours. Days. Even years. Value was measured in time.

Most transactions were digital, but physical credits still existed—rare, stamped with the faces of notable scientific and industrial figures. Because the system was tied directly to labor, inflation had become far less common. The currency itself reminded people of the value of work.

It had become a global standard. Sometimes simply called a Global.

Jake looked back at the jewelry, then at Styler, a faint gleam in his eyes.

"I'll take the necklaces," he said. "Nice and simple."

He paused.

"Four hundred hour credits." Styler considered the offer.

Four hundred hour credits wasn't bad—but he held his ground.

"I can't do less than 581," he said.

Jake raised an eyebrow.

"So… 496?"

Styler shook his head. "I really can't."

Jake paused, weighing the pieces in his hand.

"Alright," he said finally. "510. Final."

Styler kept his expression steady, hiding his excitement.

"Deal."

They completed the exchange—two aluminized necklaces for 510 hour credits. The transfer was instant. Jake sent the credits directly to Styler's holocaster, the balance updating in real time.

Styler slipped the device back into place and adjusted his backpack before stepping out of the shop.

Back on the street, he moved with a little more purpose.

As he walked, he passed another construction site. This one was staffed entirely by automated concrete pourers—unmanned units working in precise coordination, guided by satellite systems.

The satellites themselves were nearly invisible now—no larger than a fingertip, yet equipped with immense power and processing capability. Positioned farther out in orbit, they required fewer numbers and no longer interfered with space-bound vessels leaving the atmosphere.

Below, the machines poured concrete in smooth, calculated patterns, forming the base of a new structure.

A rocket base.

The building above it would be an airtight residential skyscraper, shaped like a spire.

A deliberate design. A symbol.

To aspire for more.

Styler turned a corner—and slowed. A thought crossed his mind.

Harold wouldn't like me being here.

He tightened his grip on his backpack straps. Inside, the vintage bracelet remained—still valuable.

Then he noticed them. A group of older boys.

They stepped out from the edges of the street, some emerging from the shadows between buildings. Their movements were casual, but coordinated.

Too coordinated. They circled him.

One of them reached forward, tugging at his bag.

"We're taking that." Styler's grip tightened.

Before anything could escalate, a voice cut through the tension.

"What are you doing?"

An older boy—maybe eighteen—was approaching from down the street. He wore a holocaster unlike anything Styler had seen before.

The group hesitated.

For a moment, they tried to play it off—shrugging, stepping back as if nothing had happened.

Then they dispersed.

Gone as quickly as they had appeared.

Styler exhaled.

"Thank you," he said quietly. Then, a little louder, "I'm Styler. What's your name?"

The older boy gave a small smile.

"Cody," he said. "My parents thought naming me that would make me good at coding."

He shrugged.

"Didn't hurt."

Then he added, almost casually, "Some people call me Cody Coyote."

Styler nodded, still catching his breath.

"You should've called a CopBot," Cody said.

Styler hesitated.

"What if that made things worse?" Cody studied him for a moment. "What if it already was getting worse?" There was a brief pause.

Then Cody shifted the tone.

"You seem alright," he said. "Are you doing anything in a couple of days? A friend of mine's opening a club. You should come."

Styler didn't hesitate this time.

"Yeah," he said, a little too quickly. "Sure."

For the first time that day, something felt different. Cody glanced at Styler, his holocaster already active. "Want an entrance card?" he asked.

His gaze shifted briefly to Styler's arm. "Do you mind?"

Styler nodded. "Yeah—send it."

A thin projection flashed from Cody's device. In an instant, a temporary access band formed across Styler's forearm—clean, precise, almost like it had been printed onto his skin. It displayed the date, time, and location, along with the authorization needed for entry.

Cody gave a quick nod.

"See you there." Then he was gone.

Styler stood for a moment, looking at the mark on his arm. It felt strange—but in a good way.

For the first time in a while, it felt like he had been left with something more than just another passing interaction.

Maybe even a friend. He continued walking.

His thoughts drifted upward, following his gaze into the sky. He found himself thinking about balance—about how planets might act as opposing forces to a star's pull. Some drawn inward, others holding their position through equal resistance.

Like particles. Like atoms.

The sun as a nucleus. Planets as electrons and neutrons—held in place by a force precise enough to sustain motion, but not so strong that it collapsed everything inward.

A balance between attraction and resistance.

He wondered how much of that same balance existed in the world around him—between people, systems, even the technologies they depended on.

Further along, he passed the base of a skyscraper that stretched so high it disappeared into the clouds.

Above even that, satellites operated silently—scanning structures, predicting stress points, calculating structural lean. They monitored the city constantly, identifying risks and planning future developments.

Somewhere up there, systems were already designing the next generation of space-capable buildings.

Styler slowed as something caught his eye.

Through the window of a storefront, he spotted a piece of furniture—a small, translucent table. Sleek. Lightweight. Practical.

A price flickered beside it. 80 hour credits.

He paused.

For that quality? And portable?

It felt like an opportunity. He stepped inside.

The shop was quiet. No greeting came. The owner stood nearby, occupied with something behind the counter.

Styler glanced around, then said casually, "Looks like you could use an employee." He meant it.

The owner barely looked up. "Small shop. I run it myself."

Styler nodded, pretending to browse before circling back.

"That table," he said, gesturing. "How long have you had it?"

"Since two one three six," the owner replied. "About a year."

There was a pause. The man studied him briefly—seeing a kid, not a customer.

"Look," he added, "I'm running a business here." Styler didn't hesitate. He brought up his holocaster, displaying his balance.

"So am I," he said. "Fifty hour credits for the table."

The owner blinked, caught off guard.

"Sixty-seven," he said quickly. "That's as low as I go." Styler tilted his head slightly.

"We'll do fifty-five."

The owner allowed himself a small grin.

"Sixty-three."

A beat.

Styler nodded.

"Deal."

They completed the transaction at the counter.

"Want it delivered?" the owner asked. "Or autohauled?"

Styler smirked.

"I'm a self-delivery service."

He lifted the table carefully, adjusting his grip as he turned toward the door. It was lighter than expected—but still awkward.

He knew exactly how ridiculous he must have looked walking out with it.

He did it anyway.

Styler walked with confidence—though carrying a small translucent table through the city made him look anything but subtle.

Still, he had a plan.

He made his way toward a Galleria he had only visited once before—a structure shaped like a massive glass bubble. He remembered it clearly. Inside, it specialized in high-end furniture displays.

And more importantly—

He remembered a set that was incomplete.

As he approached, he slipped into a nearby alley and carefully set the table down, hiding it out of sight. He hesitated for a moment, hoping it wouldn't be stolen.

Then he turned and entered the Galleria.

Inside, the space was bright, open, and polished. Styler approached the hostess and began speaking,

trying to explain his idea.

"I noticed one of your sets is missing a bedside table," he said. "I might have something that fits."

The hostess barely acknowledged him. Her attention drifted past him—

A couple had just entered. Styler followed her gaze.

The couple moved toward a display—a chandelier emitting a soft neon glow. The hostess stepped away from Styler and began assisting them, her tone immediately more engaged.

Styler lingered nearby, watching.

He couldn't hear the conversation clearly, but the body language was obvious.

They were interested. He saw an opportunity. Styler stepped closer.

"That chandelier," he said, addressing the couple, "has to be worth more than what they're asking."

The couple turned to him, curious.

The woman smiled slightly. "It's perfect for our dining room."

"It is," Styler agreed. "I've never seen one quite like it—and I work in the trade."

The woman raised an eyebrow. "You work at a gallery?"

Styler shrugged, a hint of humor in his voice.

"No," he said. "I work in trading." She laughed lightly.

"Well," she said, turning back to the piece, "we'll take it."

The hostess quickly finalized the sale, sending a digital holocard invoice and scheduling automated delivery for the couple.

Once they left, she turned back to Styler.

"Now that I've made a sale," she said, her tone neutral, "what were you here for?"

Styler pointed toward the furniture display.

"That translucent set," he said. "It's missing a piece—and I'm here to complete it."

The hostess looked unconvinced.

"Oh, really?"

Styler nodded.

"What if I had a table," he continued, "something that works as a bedside or corner piece—and matches the translucency of the dresser and vanity you're selling?"

He paused.

"It would turn what you have into a complete set. That increases its value."

The hostess considered this.

"I suppose it would," she admitted. "I might be interested in seeing it."

Styler smiled.

"Just a minute."

He stepped back outside and returned to the alley. The table was still there.

He quickly wiped it down, checking for any marks or dust, then lifted it and carried it back inside.

The hostess watched as he approached.

"Place it next to the set," she said. "I want to compare."

Styler carefully positioned the table beside the dresser and vanity.

For a moment, neither of them spoke. It fit.

Almost perfectly.

As if it had always been part of the set.

The hostess examined the table once more, running her hand lightly along its surface.

"Miraculous," she said. "And in good condition, too." She glanced at Styler.

"I'm interested. Since you helped me make that sale, I'll offer you 156 hour credits."

Styler didn't hesitate. "I'll take it."

The transaction was quick. The credits transferred to his holocaster, and just like that, the table was no longer his.

Outside, the city felt different.

Styler blinked, his vision struggling to adjust after the artificial lighting of the Galleria. The towering buildings blocked much of the natural light, leaving the streets dim even during the day.

He slipped his hand into his pocket and activated his holocaster, adjusting his eyesight until the blur settled into clarity.

For a brief moment, the experience stayed with him. It made him think.

There were only a few things that truly mattered—his sight, his hands, and a clear mind.

Everything else depended on those.

The holiday season was approaching.

Holidays still existed, though they had changed. Alongside traditional celebrations were new ones—dedicated to scientific breakthroughs and technological achievements. Society had shifted.

Many people now believed in a kind of scientific determinism, a structured understanding of cause and effect that governed everything.

But Styler wasn't always sure where he fit into that.

He felt lost at times—like just another person moving through a world full of people who seemed certain of their direction.

He knew he couldn't understand everything. Couldn't remember everything, even if he tried.

So maybe…

There had to be something more. Traffic moved steadily through the city.

Most vehicles were electric and fully autonomous—nothing unusual anymore. Many featured seamless glass exteriors, eliminating blind spots entirely, much like Harold's Morrison Mod 5.

Some districts had adopted alternative systems. Compost-powered vehicles, for instance, ran on processed food and horticultural waste, helping reduce the city's excess.

Fossil fuel vehicles still existed, though barely. Global reserves had nearly been exhausted. The few remaining "Fossils," as they were called, were owned by collectors with special licenses.

Styler had always wanted one. Not to drive.

To sell.

Other vehicles ran on renewable fuels—created from sources like captured carbon dioxide, converted into usable energy through advanced catalysts.

Above and around him, the city extended upward. So high that transportation had evolved with it.

Since 2130, most vehicles had been equipped to interface with rail systems built into roads, buildings, and structural supports. These car lines allowed vehicles to travel not just across the city—but up it.

Cars could ascend buildings, reach elevated districts, and recharge while in motion along these rails.

The city had become vertical.

As evening settled in, the air grew damp and heavy.

Styler made his way toward the train station, preparing for the journey back to the outskirts.

The Railway carried him out of the city, and from there, an Autocab brought him the rest of the way home.

He stepped through the door. "Supper!" Harold called out. Styler didn't stop.

"I'm not hungry," he replied, heading straight to his room.

The door closed behind him.

Inside, he checked his holocaster, reviewing his balance—counting the hour credits he had earned.

After a moment, he made a decision.

Quietly, without saying anything, he transferred a portion of his credits to Harold's account.

Then he set the device aside.

4
THE EVENT

Styler woke restlessly.

Mornings were never easy—though for him, *morning* was simply whenever he woke, no matter the hour. Fighting it always felt like a losing battle.

Today was different.

It was the day of the party.

The invitation lingered in his mind—the one Cody had given him. Styler remembered the time, the place… but not the access band. He had cleared it from his arm earlier, dismissing it as an annoyance.

Now he wasn't so sure.

Doubt crept in.

Do I even know anyone there? What if I don't fit in?

What if it's pointless?

What if I can't even get inside?

He sat with those thoughts longer than he wanted to. Then, quietly—He made a decision. He was invited.

That was enough. He would go.

To steady himself, Styler turned to something familiar. Work.

He and Harold spent part of the day reviewing schematics for the vehicle they had been building together.

Styler carefully organized the design papers across the table, aligning them as best he could.

Each page followed a system. Safety precautions at the top. Then the overview.

Then the components—numbered step by step.

Every part was drawn, labeled, measured. Materials were listed. Processes annotated. Assembly sequences mapped out with precision.

They had designed one hundred and nineteen parts so
far.

Only twenty-three had actually been built. There was still a long way to go.

As the time approached, Styler finally spoke. "Can you drive me to the station?" he asked. Harold glanced up.

"Of course," he said. "I'll drop you off on my way to work."

They drove out together in the Morrison Mod 5, taking the ocean-view route toward the rail station.

The world felt more connected than ever in 2137.

Beneath the surface of the water, high-speed trains moved along submerged tracks near the ocean floor—vast underwater rail systems linking entire continents. While passengers could use them, they were primarily designed for freight.

Through the vehicle's clear panels, Styler caught glimpses of them—long, shadowed forms cutting through the water, leaving trails of movement behind. Cargo trains, likely bound for the city.

Harold glanced over briefly.

"What've you got planned for tonight?" he asked.

Styler answered without hesitation.

"Nothing."

Harold frowned slightly.

"I'm driving you to the rail station for nothing?" Styler allowed himself a small smile.

"How about this," he said. "I'll tell you when I get back."

Harold chuckled.

"Now you're trying to sell me?"

Harold drove Styler to the station, glancing at him now

and then—still quietly wondering what he was up to. "I want you to be safe, alright?" Harold said. Styler nodded.

They pulled up to the station.

"Alright," Harold added, "here we are. I'll leave you to it."

Styler stepped out, giving a quick wave before heading inside. He made it onto the rail just in time.

Inside the train, he moved down the aisle, scanning for a place to sit. Out of instinct—part caution, part habit—he chose a seat near a pair of passengers who looked close to his age. Not too close. Just enough to feel… safer.

The ride was quick.

When he arrived, he stepped off and made his way toward the

venue, not far from the central station.

Above him, autonomous helicopter taxis moved through the city—small drone-like cabs weaving between buildings, even passing through designated openings carved into the structures themselves.

Styler watched them briefly. He could never afford one. So he walked.

By the time he arrived, dusk had passed. It was nearing eight.

The venue stood at the base of a towering skyscraper, its exterior surrounded by monumental statues. Neon light flowed across them—shifting colors projected through, around, and within the structures, giving them an almost living presence.

A crowd had already gathered.

Hundreds of people stood in line, adjusting their appearance, checking reflections, preparing to be seen.

Styler moved toward the line, brushing past someone without getting a proper look.

At the entrance stood something unusual. A barrier.

Thick—almost like reinforced glass. They called it *the walk-through.*

One by one, people stepped forward…

…and passed straight through it. As if the solid surface wasn't there.

Styler watched carefully, trying to understand. Then he took his place in line.

It didn't take long for a realization to set in. He didn't have a wristband.

He exhaled slowly, thinking. Then—an idea.

Subtly, he pulled out his holocaster and angled it behind him, zooming in on the wrist of the person standing just back in line. A projection band—just like the one Cody had given him.

The one he had erased.

Carefully, he activated a custom scan.

Copy.

Replicate.

A faint projection formed across his own arm. He slipped the holocaster away.

Now here I am, he thought. *Sneaking into something I might not belong in.*

His turn came.

Styler stepped forward toward the walk-through. For a brief second, hesitation hit him.

What if it doesn't work? What if I just slam into it? He took another step. And passed through.

The sensation surprised him—not resistance, but relief. A soft, almost therapeutic feeling washed over him as he moved through the barrier.

No impact.

No rejection.

Just passage.

He was inside.

Behind him, the line continued to move.

A group of boys stepped forward, energized, laughing as they approached their turn.

A few made it through. Then—one stopped.

The same boy Styler had scanned from. He walked straight into the barrier—And bounced back.

"What the—?!"

The surface flashed red. Denied.

He tried again.

Still nothing.

His friends looked back, confused, but were already being pulled inside by the flow of the crowd.

Styler stood just beyond the threshold. Inside, the environment shifted immediately.

A glowing display hovered in the air—text arranged like a structured script, written in precise, technical formatting. Beneath it floated a holographic interface, designed for hand-scan authorization.

A policy.

Clear.

Enforced.

Follow the rules—or face the consequences. The system waited.

And it read…

The glowing policy hovered in the air, written in structured lines—almost like code.

It read like a philosophy. Or a warning.

Relationships, it suggested, should be approached without emotional bias. Not driven by impulse, attraction, or habit—but by evaluation. By evidence. By outcomes.

Compare the past to the present. Assess usefulness.

Search through holocasters or other systems for compatibility.

Ask questions that matter—family, stability, contribution.

If there are no attachments, perhaps they are suitable. Feelings, it implied, were distractions.

You could feel them—but why should you trust them?

The message continued beyond what Styler could absorb at once, its logic stretching into something both clinical and unsettling.

A name floated at the top:

"Better Good Dating Agreement."

Styler leaned in, almost pressing his face toward the glowing text as he read. He had never been inside a place like this before. It felt... older than him. Designed for people who understood things he didn't yet.

Still—He raised his hand and signed.

Nearby, an older boy read through the agreement carefully.

"That's actually useful," he muttered.

Someone passing by responded without slowing, "That's right."

A club employee guided a group inside.

"Friend party of yellow—walking in yellow!" One of them laughed.

"I guess that means you should always say hello," he said. "Anyone wearing yellow is the most important person to meet."

It sounded welcoming. Almost too welcoming.

Like a rule that carried weight beneath its surface. Inside, the club came alive.

Holographic dancers moved across elevated platforms, their forms shifting with impossible precision. Laser lights cut through the air while AI-generated music pulsed through the space.

Above, holograms released streams of confetti that dissolved before ever touching the ground.

Styler watched closely.

The detail was astonishing—almost too perfect. For a moment, he wondered if the developers had ever feared the strain such visuals could put on the human eye.

Everything felt intense. Overwhelming.

He moved through the space slowly, unsure of where to go.

A simulated storm drifted across the ceiling—dark clouds forming, shifting.

Then came the rain.

Red.

Holographic droplets fell in patterns, forming shapes mid-air before dissolving—never touching the people below.

Styler stood beneath it, feeling out of place. Alone.

"Styler!"

The voice cut through the noise.

Cody approached quickly, wearing a dark slim-fit jacket and blue jeans. His hair shimmered silver—shifting tones every few moments from a reactive gel.

"Glad you made it," he said. "Come on—I'll introduce you to some people."

He guided Styler toward a group gathered off to the side.

"Hey, we've got someone new," Cody said.

He gestured one by one.

"This is Gavin. Rex. And Tim. Guys—this is Styler." Styler gave a small nod.

"What's going on?"

Gavin grinned immediately.

"Eleven-zero season," he said. "That's what's going on."

Without pause, he launched into a story about his father's team—the HellCats—detailing wins, rankings, and expectations like it was all second nature.

Rex followed, shifting the conversation to his own background—his family's past ownership of a major fossil fuel car company before the energy era changed everything.

Styler listened.

Quietly.

Their histories sounded complete. Defined.

He realized how little he knew about his own. Tim glanced at him.

"You sign the Better Good agreement yet?" Before Styler could answer, Gavin smirked.

"The better good," he said, "is no one ever dating you."

The group laughed.

Styler forced a small smile—Still trying to find his place.

Styler realized something quickly.

Cody was the only reason he felt even remotely comfortable there.

"I'll see you around," he said, stepping away from the group.

He moved into the crowd, letting himself disappear into the shifting sea of people. Behind him, Cody's hair had already changed again—now a deep, lustrous purple under the club lights.

As Styler wandered, his attention shifted upward. On the second tier, he noticed her.

A girl standing alone.

She wore holocaster glasses—sleek, mostly holographic, with subtle golden accents that framed her face perfectly. She held a drink in one hand, her posture calm, self-contained.

Something about her stood out. Familiar.

Styler made his way up the clear glass steps, each step catching and reflecting the lights around him. The elevated platform gave a clearer view of the club below—but his focus remained fixed on her.

He approached carefully.

The noise was loud—constant—so he leaned slightly closer, speaking near her ear.

"Everything you can't see through is a door," he said softly. "Your closet. Your dresser."

A brief pause.

"I want to see through you… just for a moment." He

straightened slightly.

"My name's Styler. What's yours?" She turned toward him.

"Kailee."

He nodded, almost relieved.

"It's just…" he added, searching for the right words, "something about you reminded me of myself."

Across the main floor, movement shifted. The club manager was making his rounds.

He moved with purpose, wearing a specialized electro-embedded armband integrated into his clothing. It glowed subtly, functioning far beyond standard devices—scanning, processing, verifying.

As he passed through the crowd, he checked wrists. Authorization.

Access.

Then deeper scans—detecting anything out of place. Styler noticed immediately.

A sharp tension ran through him.

What happens if he scans me?

The manager's attention shifted.

Toward him.

Styler stepped back slightly, his voice low and quick.

"I have to go," he said to Kailee.

Then he moved.

He slipped into the crowd, weaving between people, keeping his head low. Every movement was deliberate—avoiding direct lines of sight, staying unpredictable.

Behind him, the manager continued forward. Closing distance.

Styler spotted an exit toward the rear.

He pushed through, reaching the far corner of the space—

But it wasn't an exit.

It opened into a mag-lev elevator access point. A shaft.

No time to hesitate.

He pulled out his holocaster and activated it, recalling fragments of code Harold had taught him—just enough to bypass basic restrictions.

He focused.

Thought.

Executed.

The door responded. Unlocked.

A clear path opened. Styler stepped forward—

And slipped inside.

Styler's first thought, standing in the elevator shaft, was simple:

Someone had to have died building this.

The space was dark, narrow, and industrial—far removed from the polished illusion of the club above.

He found the service ladder and began climbing. It held his weight easily.

Step by step, he pulled himself upward, the metal rungs cold beneath his hands.

Not only do you have to prove yourself, he thought, you have to prove yourself your entire life.

The thought stayed with him as he climbed.

He reached the next level and forced the elevator doors open just enough to pull himself through.

The second floor.

Quieter.

Dimly lit.

This part of the building felt different—less spectacle, more function. Offices. Storage. Administrative space.

Styler moved quickly, scanning for an exit. A staircase.

A fire door.

Anything.

He found one.

Moments later, he was back outside.

The air hit him differently this time.

He didn't stop.

He kept moving—fast at first, breaking into a run, putting distance between himself and the club. After a few blocks, his pace slowed to a steady walk, his breathing catching up with him.

He was in the clear. Still—

He kept going.

Up ahead, a crowd had gathered. There was noise—but not

panic. Excitement.

Cheers.

Styler slowed, watching.

People were pressing inward, voices overlapping—"John! John!"

"Can I get your signature?!"

Styler stepped closer, blending into the crowd, curiosity pulling him in. He moved carefully, unnoticed, slipping between people until he reached the front.

Then he saw him. John Williamson.

Recognition hit immediately.

His teacher would never believe this. Styler took a step forward.

"Mr. Williamson," he said, his tone steady despite the moment, "I'm a fan."

He hesitated only briefly.

"I invent too," he added. "With my grandfather." John Williamson looked at him—really looked.

Then smiled.

"Keep going, kid," he said. "You're my kind of person."

The words landed harder than Styler expected.

Before he could second-guess himself, he spoke again.

"Would you take me as an understudy?" It was direct. Honest.

In a world built on systems and access, why not ask?

Williamson glanced toward someone nearby.

"Give your name and imprint to my publicist," he said. "I'll get back to you."

The publicist stepped forward, quickly recording Styler's information.

Styler couldn't hold it in.

"Thank you, Mr. Williamson!" he called out, his voice cutting through the noise.

Moments later, he stepped back from the crowd. The city felt different again.

Faster.

Full of possibility.

He turned toward the station and made his way back to the rail—ready to get home.

The next day, Styler decided to spend his Sunday in the city.

Harold had the day off, so Styler took the Morrison's

Mod 5 using its autonomous driving feature. Even with self-driving systems, regulations were strict—every passenger was required to carry their digital imprint at all times.

Identification had evolved.

Driver's licenses, birth records—everything was now tied to a person's DNA. Full genetic sequencing was required for official documentation. With the global population at such a scale, physical resemblance between individuals had become too common.

Holocasters handled it all.

They weren't just tools—they were identification, security, and access devices. With near-infinite battery life, they had become something people were legally required to carry at all times.

On the way, Harold spoke casually, as he often did when it came to practical advice.

"When you're older, you might need your own car," he said. "Best place to buy one? Right after a factory tour. You learn how it's built—maybe even how to fix it."

He smiled slightly.

"Unless you build your own. That works too… as long as it's road legal."

A brief pause.

"Inventing your own things—that's what makes people useful."

He continued, almost as an afterthought,

"Same idea with food. Best place to get it? A farm.

Especially if you learn how to grow it yourself." They reached the station.

Styler stepped out, boarded the rail, and before long, he was back in the city.

Above him, a broadcast flickered across the sky.

The projection was unstable—tilted at an odd angle, its visuals glitching. The sound came through distorted, breaking in and out.

Damage from a lightning strike days earlier. Still, it continued to transmit:

"AI scanning people left and right… nothing new. If you don't have a profile, one will be made for you. AI scan for access."

The message repeated, fractured but persistent. Styler made his way to his usual pawn shop.

Inside, everything was as he remembered. The perpetual motion device on the counter continued its endless rotation, humming softly in the background.

Jake stood behind the counter, waiting.

"Hey, Styler!" he said with a grin. "How's it going?" Styler sighed slightly, pulling out the bracelet.

"Still trying to get rid of this."

Jake glanced at it, then shook his head. "Not right now. Inventory's stacked." He paused, thinking.

Then—"I've got something else," he said. "Small job."

He reached behind the counter and set down a box.

"Take these vintage light bulbs over to Fifth Street Pawn. I'm clearing stock—selling to competitors."

He looked at Styler more seriously.

"I'll give you some hour credits for your time." Styler nodded.

"Sure thing."

Jake's tone sharpened just slightly.

"I want these handled carefully. They're already paid for."

Styler carefully lifted the box of vintage light bulbs Jake had pointed out. He already knew the destination—Fifth Street Pawn—so he stepped out of the shop and headed in that direction.

As he walked, a thought began to form.

What if I did this full time?

A delivery service. His own business.

Something simple to start—something that could grow.

But then he looked down at the box in his hands. The weight of it felt heavier than it should have. *No…*

That wasn't what he wanted.

It felt like settling.

Like drifting away from something bigger—away from what he and Harold were building. Away from the ideas he hadn't fully realized yet.

He kept walking.

Fifth Street Pawn came into view.

Its signage glowed brightly—almost excessively so—while the exterior itself looked worn, like it hadn't been updated in years.

The doors slid open as he approached.

Inside, the shopkeeper looked up immediately.

"You look like you've got an idea!" he said, pointing at the box. "Light bulbs—get it? Idea? Come on, that's funny."

Styler stared at him blankly. The shopkeeper shrugged.

"Alright, tough crowd. I've been waiting for those, by the way."

Styler hesitated for a moment. Then, almost instinctively, he spoke. "I actually do have an idea."

The shopkeeper paused.

"Oh yeah? Let's hear it."

Styler adjusted his grip on the box.

"What if light didn't need a source?" he said. "No bulb.

No fixture. Just… light. Existing on its own." The shopkeeper frowned slightly.

"How would that even work?"

"You said 'idea,' right?" Styler continued. "Think about it—air is just there. The space we move through—it's just there. What if light could exist the same way?"

The shopkeeper leaned in a little.

"Explain that."

Styler nodded, finding his rhythm.

"Light that floats in a room. You can place it anywhere. Move it. Shape it. No replacement, no repairs. Just… light, existing by itself."

The shopkeeper's expression shifted.

"Now that," he said slowly, "makes more sense. That would save me a lot on fixtures alone."

Styler smiled slightly.

"We could build it," he said. "You and me. Something like… 'nothing light.' Or… 'light from nothing.'"

He paused.

"We'd figure out the name."

The shopkeeper shook his head, though not dismissively.

"I like the idea," he said. "I do."

Then his tone changed—quieter, heavier.

"But I just run a shop." He looked around briefly.

"You're young. You can still change things." A beat.

"I've already spent enough years figuring out I didn't." The words settled between them.

Then he reached forward and took the box from Styler's hands.

"I'll take those."

The weight was gone.

Styler stepped back slightly, watching as the shopkeeper set the box aside.

"Thanks for delivering them," the man added.

Styler nodded.

Outside again, Styler paused for a moment. Then turned.

He decided to head back toward Jake's shop—partly for the credits he'd been promised, but also because something about the conversation lingered.

This time, he took a different route.

Walking along Avenue Terrace, Styler glanced through the glass of Fill—a small, brightly lit beverage spot known for its futuristic drinks.

That's when he saw her.

Kailee.

She sat alone by the window, her golden holocaster glasses faintly shimmering as she read. In her hands was something rare—an old paperback book. A relic. The kind that barely existed outside museums or private collections anymore.

Styler slowed. Then stopped.

Fill specialized in engineered beverages—slightly fizzy liquids that solidified once consumed, creating a lasting sense of fullness. With customizable flavors and adjustable duration, a single drink could replace meals for hours… even years.

The cups themselves were designed with a curved pocket, mimicking the upper shape of the stomach.

Styler barely noticed any of that right now. His attention was fixed on her.

He stepped inside.

She didn't look up.

For a moment, he just stood there, unsure. Then—an idea.

He approached the self-service station, holding a full cup of Fill. Without hesitation, he emptied it discreetly into a nearby disposal unit.

The cup remained.

He took something sharp and etched a message into its surface—carefully, quickly.

Then he grabbed a handful of stirring sticks from the counter.

Straws, really. Always straws. He walked over.

Kailee looked up, clearly annoyed at the interruption—then paused.

Recognition.

"You're familiar," she said. "I met you… last night, didn't I?"

"Kailee," Styler replied. She nodded. "Yes."

Styler smiled, placing the stirring sticks gently on the table as he sat.

"You and me," he said. "Let's play a game." He gestured to the sticks.

"One question per straw. Until we run out."

He set the cup down in front of them—carefully positioned so she couldn't see what he had written.

Kailee considered it—Then smiled.

"Alright," she said. "I'm in." "Go ahead," Styler said.

"I'll start," she replied. "What's your ancestry?" Styler dropped a straw into the cup.

"Arlenish," he said. "You?"

"My family traces back to Gerdeen," she answered. "But we've been here a long time."

She added a straw.

Styler picked one up, twirling it slightly.

"These make good bookmarks," he said.

Then—"What does your family do?" Another straw into the cup.

Kailee leaned back slightly, thinking.

"My father's a politician," she said. "Corporate events, public functions—we go to a lot of things."

A small smile.

"My mother raised me and my sister, Constance. She's involved in community work."

She dropped in another straw.

"And you?"

Styler hesitated. Just for a moment.

Something quiet passed through him.

"My grandfather works at Local Tech," he said. "He's a technician."

Kailee nodded—but there was a flicker of curiosity behind it.

A question she chose not to ask.

"Any siblings?" she said instead, adding another straw.

Styler smiled.

"No. Just me."

The questions continued. Simple.

Unexpected. Sometimes playful. Sometimes thoughtful.

Straw by straw, the cup filled. Eventually—they ran out.

Styler paused.

Then slowly turned the cup around. The etched message caught the light. **Kailee, go out with me? :)**

For a second, she just looked at it. Then she smiled.

A real one.

"I'll cast you," she said. Styler nodded, standing. "I'm counting on that." He stepped back slightly.

"I'll let you get back to your book."

As he walked out through the automatic doors, something lingered.

A feeling.

Quiet, but certain. Something had changed.

5

A BOOK FOR STYLER

Styler sat in his secondary school classroom, his holocaster resting on the desk in front of him.

It was exam day. And he felt it.

A quiet tension sat in his chest—steady, persistent. The teacher stepped forward.

"Everyone settle down," he said. "Take out your holocasters."

The room shifted as students adjusted in their seats, activating their devices.

"We're about to begin your year's comprehensive," the teacher continued. "Mathematics, biology, physics… all of it."

He paused, scanning the room.

"I'm confident you're all prepared."

Sean leaned toward his friends, muttering under his breath.

"What planet is he on? Beta 16.7?" he said. "You all feeling prepared?"

They exchanged blank looks.

"Enough," the teacher said sharply. "I heard that." The room quieted.

"Remember this," he added. "Nothing is hard. And nothing is easy."

A beat.

"Holocasters ready."

"I'll be queuing the exam in two minutes," he said. Then, without looking directly at him—"And no, Sean, you won't be able to cheat."

Sean tapped his fist lightly against the desk, frustrated. Michael raised a hand slightly from a few seats over. "Are we ever actually going to use math?" he asked.

The teacher didn't hesitate.

"Look around you," he said. "Everything you see—everything you touch—exists because of math and science."

He stepped forward.

"Every object you've ever used. Every system you rely on. Every purchase you've made."

A pause.

"That's all math." He let that settle.

"Be good at it, and you could become the next great inventor."

Another pause.

"The next John Williamson." The name carried weight.

A pioneer in outer-quadrant space technology—though not without controversy. Some believed his work was exaggerated. Others questioned whether figures like him were even entirely real anymore.

But the idea remained. Possibility.

The teacher's tone softened. "Everyone settled?"

He looked across the room.

"Before you begin, think about this."

"What matters more—getting the easy questions right, or solving the hardest ones?"

Silence.

"How do you feel when you solve something easy?" A few students shifted.

"And how do you feel when you solve something difficult?"

"The easy questions make you confident," he said. "But they don't define you. Everyone can answer those."

He gestured lightly.

"Skill comes from the hard ones." His voice steadied.

"In life, the hardest problems are the ones that matter.

The ones worth solving."

He stepped back.

"This exam has no multiple choice." A few students reacted quietly.

"Life doesn't give you options to pick from," he continued. "It doesn't hand you answers."

A final pause.

"You have to create them." He nodded once.

"Begin."

After class, Styler didn't check his results. Pass or fail—it didn't matter. Not right now. Instead, he took his time heading into the city.

The streets were crowded, alive with movement. Buildings rose not just upward, but outward stacked, layered, pressed tightly against one another. The city felt dense. Compressed.

Styler walked through it all, observantly. Curious.

Detached.

He passed a row of university buildings; each marked with bold signage:

Engineering.

Architecture.

Chemistry.

Geology.

Oceanography.

Styler slowed.

Then he laughed quietly to himself.

What kind of joke is this?

His thoughts began to race. *Oceanography? In the middle of a city? Chemistry—where? There are no factories here. Engineering… with what materials?*

Geology? There's no earth left to study—just concrete.

He glanced around.

Agriculture? There isn't a single farm in sight.

To him, it didn't make sense.

Not fully.

It felt disconnected.

Like learning without substance.

Styler had been to the city many times, but it never inspired him.

If anything, it unsettled him.

He couldn't see a future for himself here.

Or for his family.

In a place like this, he thought, even people who worked weren't truly secure. You didn't have to be homeless to feel like you were struggling just to exist.

Sometimes it felt like everyone was asking for something.

Even those with jobs. The city didn't produce. It consumed.

Everything—materials, food, energy—came from somewhere else. From rural regions. From outskirts like his own.

And yet—the city kept growing.

Styler watched as construction continued around him. Robotics handled most of the labor, but people were still there—guiding, assisting, building.

Building structures, they would likely never live in. Never afford.

That thought stayed with him. He imagined it.

Spending your life constructing something you would never belong to.

A future you help create—But can never enter.

In places like this, resources didn't exist in their raw form. There was nothing to extract, nothing to shape from the ground up. Everything was imported, assembled, distributed.

For someone like Styler—someone who wanted to build, to invent—it felt limiting.

Some planets had already outlawed cities like this.

Places that couldn't sustain themselves.

Places that depended entirely on external systems to survive.

Styler looked around once more. So many people.

So much movement. So much noise.

And still—a question lingered.

Does anyone here truly belong?

There was something else Styler couldn't ignore about the city.

The absence of smell.

It bothered him more than he expected.

In crowded places—transit stations, packed streets, long lines of people pressed shoulder to shoulder—there should have been something. Sweat. Fragrance. The natural presence of people.

But there was nothing. No scent at all.

It felt wrong.

Artificial.

Almost sterile in a way that made him uneasy. He pushed the thought aside.

His mind shifted—to something quieter. More personal.

The future.

He found himself thinking about Stout University.

A place he had only ever heard about—almost like something out of a storybook. It fascinated him.

There were assembly plants nearby. Hands-on learning. Real machinery.

Classes that went beyond walls—into the outdoors—teaching students how to work with nature instead of just studying it.

That felt real to him. Useful.

But then the question surfaced.

Should you pay for something intangible like education?

He frowned slightly.

What if you didn't?

What if you negotiated it?

People negotiated everything else—cars, deals, contracts.

Why not *careers*?

Why not walk in and make a case?

They had business departments, didn't they?

Then treat it like business. The idea lingered.

He knew one thing for certain—to build something meaningful, you couldn't stay in one place.

Resources were scattered. Opportunities were scattered. If you wanted to create, to produce, to innovate—you had to move.

Constantly.

It didn't matter what you were aiming for.

Movement was part of it. Still—

Was school even necessary anymore? Technical skill mattered most.

Yes, institutions had equipment—advanced, valuable—but you didn't own it. You couldn't take it with you.

So was it worth it?

As he walked, his thoughts were interrupted. Graffiti.

Scrawled across the walls of an occupied building. The markings were in Nadjin—uneven, deliberate. Styler couldn't read it.

But he understood the tone. A warning.

Stay away.

He kept moving.

Soon, he entered another district.

The atmosphere shifted immediately.

This was Time Travel Avenue.

Both sides of the street were lined with commercial time travel venues—each louder, brighter, more attention-grabbing than the last.

Discount Time Travel.

Solar T Travel.

Medieval Time Travel.

Primo Time Travel.

Your Pick Time Travel.

Animatronic displays spilled out onto the sidewalks, calling to passersby. Some venues advertised precision. Others emphasized experience. Some relied on spectacle alone.

There were guides, too—cybernetically enhanced—offering curated journeys through different eras.

The entire avenue pulsed with possibility. And noise.

Time travel had become accessible.

Common, even.

But not everyone chose to use it.

Some people understood the risks—what it meant to step outside your own time, even temporarily.

Some things weren't meant to be revisited.

Styler slowed.

Then stopped.

Across the street—He saw her.

Kailee.

Standing just outside the flow of the avenue.

For a moment, he hesitated.

Should I go over?

A beat. Then—He did.

"Kailee!" Styler called out as he approached. "Hi… why haven't you contacted me?"

Kailee hesitated, a faint blush rising to her face.

"Well…"

Styler caught it immediately—the hesitation, the discomfort—and shifted the conversation.

"So… Time Travel Avenue?" he said, glancing around. "You like this kind of thing?"

"I've been before," she said quickly, brightening. "I actually liked it."

Styler nodded.

"I haven't," he admitted. "First time in this district.

Honestly… it kind of scares me." Kailee smiled lightly.

"I live nearby."

But something about her felt distant—like her attention was somewhere else entirely.

"I've just been busy," she added, almost apologetically. "I'm actually on my way somewhere right now. But—I do want to talk to you again."

Styler studied her for a moment, then nodded.

"Alright," he said. "I'll leave you to it. Cast me whenever."

Kailee smiled, stepping away. After a few steps, she glanced back.

Styler was still standing there.

He smiled back.

Then his attention drifted.

Time Travel Avenue.

He had never been inside one of the venues before.

Curiosity pulled at him.

He approached the translucent entrance of #1 Time Travel and stepped inside.

A live attendant sat behind the counter, staring off as if lost in thought.

"It's sad," the attendant murmured to a coworker. "That old woman—she comes every day. Goes back to see her son before he died…"

Styler paused, having overheard. He stepped forward.

"Anyone can time travel?" he asked. "Anytime they

want?"

The attendant looked at him.

"Anytime," he said. Styler raised an eyebrow. "Anytime?"

The attendant smiled.

"Anytime… between 8 a.m. and 10 p.m." A brief beat.

"Just load your hour credits, put on a uniform, and you're good to go."

Styler nodded slowly.

"I'm not going," he said. "Just curious." The attendant gave a small nod.

"Anytime."

Styler stepped back out onto the street.

Not far away, Kailee lingered in the shadows.

She hadn't gone far.

Hidden just around the corner near a closed library museum, she pulled her jacket over her head—shielding herself from scanners, cameras, anything that might track her presence.

She watched.

Waited.

Then she saw her moment.

An employee exited the building.

The heavy, archaic door began to close behind them, guided by an old pneumatic arm.

Kailee moved quickly.

She slipped forward, caught the door just before it sealed, and squeezed inside.

She was in.

Inside, the museum felt still. Quiet.

Rows of preserved anthologies lined the space—carefully maintained, though here and there a worn, aging book broke the illusion of perfection.

Kailee moved through the aisles, her gaze shifting across the shelves.

She passed the science fiction section and paused briefly.

It almost felt like a game those authors must have played—who could predict the future best? Who could write it into existence?

Now, most of those books were relics. Deforestation had changed everything.

Printed material had been restricted—largely replaced by digital media. Only technical manuals, scientific texts, and essential documentation were still produced in physical form.

The rest—Left behind.

Kailee moved quietly through the museum, her jacket still pulled low over her head as she made her way toward the history and culture section.

She scanned the shelves quickly—Then found it.

The book she had been looking for. A book for Styler.

She held it carefully, almost protectively, before turning and making her way back toward the exit. This time, smoother. Faster.

She slipped out into the city unnoticed. A sudden voice cut through the air. "Welcome!"

Kailee flinched.

A fanless drone hovered directly in front of her, projecting a soft holographic display as it spoke.

Her heart jumped—just for a moment.

Did it see me?

She steadied herself.

"Thank you," she said calmly.

"Do you require any assistance in your travels?" the drone asked.

Kailee smiled, tightening her grip on the book.

"No, I'm fine."

She gestured casually toward a couple walking nearby.

"Maybe they need help." The drone paused. "Right away."

It zipped off toward them.

A moment later, she heard it again in the distance—"Welcome! Do you require any assistance in your travels?"

The exact same tone. Unchanged.

Kailee allowed herself a small breath of relief. Snow began to fall.

Soft at first—light flurries drifting between the towering buildings. The snow collected along the wide sidewalks, cushioned with recycled rubber designed to soften every step.

The city shifted with it. Quieted.

For Styler, somewhere else in the city, the sight carried something deeper.

A feeling.

Almost like Christmas. But not quite.

In 2137, real holiday plants were gone.

Outlawed.

Only synthetic, reusable versions remained.

Neither Styler nor Kailee had ever seen—or even smelled—a real Christmas tree.

A vibration broke the moment.

Styler reached into his pocket, unfolding his holocaster and activating it.

A message. Kailee.

"Are you still in the city?" she asked. "Let's go out to dinner. I know a place you'll like."

There was a pause—then another message.

"There's something I have for you. Talking to you earlier… it made me think of it. When I met you, I knew I had to get it for you."

Styler smiled slightly.

"Yeah," he replied. "Let's meet."

Later, they stood together outside a restaurant in a newer district of the city—closer to the water.

This part of the skyline felt different.

Less rigid.

Buildings here carried color, texture—variation.

A break from the monotony of older sections.

The restaurant sat at the base of a towering skyscraper.

Above it—something remarkable.

A vertical greenhouse system stretched upward, level after level of translucent structures stacked toward the sky. At the very top, a wide, flat greenhouse capped the building—resting like a crown.

Each level grew something different. Vegetables.

Grains.

Fruits.

Even full-sized trees.

Everything was automated.

A continuous conveyance system moved produce downward through processing levels before distributing it—feeding the restaurant below and supplying other parts of the city.

The flat design at the top wasn't just aesthetic.

It made maintenance easier. Safer.

More efficient than the angled towers of older designs. Styler looked up at it all.

For once—the city didn't feel entirely disconnected from something real.

Inside the restaurant, Styler and Kailee sat across from one another at a hovering round table—its smooth white surface inset with subtle grooves for plates and utensils. The table floated effortlessly, unsupported, as if suspended in air. Some said it was magnetic repulsion. Others weren't sure at all.

Above them, a network of conveyors moved silently.

Meals arrived from overhead—lowered precisely onto each table without the need for servers.

Their first course descended at the same time. Two identical portions.

Small, cube-shaped entrées—freeze-dried and delicately plated.

Styler lifted his, glancing at Kailee with a faint smile.

"I could feed you, if you want."

Kailee smiled back, already mid-bite, amused but saying nothing.

As they ate, Styler leaned in slightly, his voice quieter now more personal.

He spoke about his ideas.

About what he and Harold were building. About what could exist.

"What if it's all real?" he said. "What if one day I could build it—factories, systems… everything?"

He paused, thinking aloud.

"What if I could just… source parts from anywhere? Bring in materials, build something bigger than where I am?"

Another dish arrived.

Soft.

Melted.

Almost liquid.

They reached for it at the same time, catching each other's eyes as they did.

A brief, shared moment. The conversation shifted.

Styler found himself talking about his childhood—about growing up with only his grandfather.

He smiled as he spoke.

"I used to cut my own hair," he said. "Felt like I was preparing for a space mission every time."

A small laugh.

"It just took me forever to figure out the back." Kailee smiled.

There was something genuine in it. Something she found… endearing.

A quiet settled between them.

Not uncomfortable.

Just still.

Kailee took the moment.

She reached forward and placed something gently in front of him.

The book.

Styler looked down at it, surprised.

He picked it up carefully, almost instinctively treating it like something fragile—something important.

The plates cleared themselves silently.

Above them, a final dish hovered into place—an orb of food, softly suspended in the air. Its colors shifted naturally, reflecting its ingredients.

It needed no plate.

But Styler's attention was elsewhere.

The book.

He turned it over in his hands, studying it. Then opened it.

The pages were worn.

Delicate.

He leaned closer, almost breathing it in. The title:

"Cultures from This Planet."

On the cover, a globe—rendered with quiet authority. He flipped through.

Stopped.

Page sixty-seven—Culture: Arlenish

Arlenish are an island culture whose foundation was invention and discovery. Their contributions were so significant they helped drive the fifth wave of industrialization, including the energy revolution.

Population: 48 million (2080), 23 million (2110)

Status: Severely at Risk

Projected Decline: Continuing Styler read in silence.

Then continued more pages.

More fragments of history.

More truth than he had ever known about himself.

Something shifted.

The numbers stayed with him. The realization settled slowly—then all at once.

His culture.

Fading.

Diminishing.

Close to disappearing.

For a moment, it felt overwhelming.

Like standing in the aftermath of something lost. Something he had never even fully known.

He didn't speak.

Didn't move.

His thoughts dimmed—quiet, heavy. Across from him, Kailee watched.

Patient. Unsure. Then—"Styler!"

Her voice broke through. Bright. Hopeful.

"Did you like it?" He looked up.

And smiled.

"I love it."

The words came easily.

Even if the feeling behind them was more complicated.

He knew she hadn't meant to hurt him.

He just didn't have the language yet to explain what it had stirred.

Kailee watched him carefully.

"I wanted it to inspire you," she said softly.

Styler nodded.

"It does," he replied. "It just… also made me feel alone."

Kailee leaned forward slightly.

"It wasn't meant to make you sad," she said. "I remembered that it had information about different cultures—I didn't know exactly what was in it. I just thought… you should know your history."

A pause.

"There were things in there I felt like you deserved to know."

Styler understood.

There was no harm in her intention.

Only truth.

And sometimes, truth carried weight.

They continued talking as they ate, the conversation easing back into something lighter. By the end of the meal, they were both full—comfortably, almost overwhelmingly so.

Kailee gestured toward the bill.

"I've got it."

Styler hesitated, then nodded.

"Thank you."

They stood.

Faced each other.

A quiet moment before parting.

Kailee turned, beginning to walk away.

"I'll always miss you…" she said lightly, her voice warm despite the heaviness of the evening.

Styler gave a small, respectful bow.

"And I'll miss you too."

They both walked in opposite directions. Still glancing back.

Still holding onto the moment.

I'll miss you… until then, Styler thought.

As he walked, the weight returned.

The book in his hands felt heavier now.

His thoughts circled back to what he had read. His parents.

His past.

His culture.

He stopped.

His eyes stung.

Then—he broke.

Tears came quietly at first. Then fully.

Unstoppable.

He opened the book again, his hands trembling slightly, and without thinking, tore a page free.

He used it to wipe his eyes.

The ink blurred as the tears soaked through. For the first time—he felt it completely.

He wasn't just without parents.

He wasn't just searching for belonging.

He was part of something that was disappearing.

And now—he knew.

The Next Day.

Styler sat in class, quieter than usual.

Still.

Distant.

It was one of those slow days—the kind where students drifted, half-awake, waiting for time to pass.

Then

"The exam results are back," the teacher announced. "They're on your holocasters."

The room shifted instantly.

Devices activated.

Whispers spread.

"Styler!" someone called out.

"You got the top score!"

A ripple moved through the class.

He hadn't checked.

But it was true.

The teacher nodded.

"Everyone did well," he said. "I'm proud of all of you." Then—

"Styler, stay after class." The room emptied.

Styler lingered, absently folding and unfolding his holocaster in his hands.

"Styler," the teacher said, drawing his attention. "Have you thought about what comes after secondary school? University?"

Styler didn't answer immediately.

The teacher stepped closer.

"These scores… they're exceptional," he said. "Highest in the region."

A pause.

"Let me give you some advice." "Forget degrees."

"Forget credentials."

"Forget authority—even mine."

"They don't mean what people think they mean," he continued. "They're often just… structures. Titles. People playing roles."

He held Styler's gaze.

"If you follow titles instead of people, you'll be disappointed."

"Some of the smartest individuals never had degrees," he said. "They figured things out on their own."

A small nod.

"That's real understanding."

"If you go to university," he added, "don't pay for it.

Or don't go at all."

Then, more thoughtfully—

"But I know you. You like to build. To invent." "Engineering might suit you."

A beat.

"But don't just follow a system." "Study the exams themselves."

"Study the questions." "Break them down."

"Reverse engineer the knowledge." "Find your own answers."

The teacher continued, his tone shifting—more grounded now, more practical.

"If you do go to university," he said, "don't think of it as studying one subject in isolation."

He paced slowly.

"Engineering? You won't build anything meaningful alone."

"Chemistry? There's no such thing as chemistry by yourself."

A pause.

"Nothing that exists in the real world—no product, no system, no structure—is created by one person."

He looked directly at Styler.

"To bring anything into production, it takes people. Dozens. Hundreds."

"Every class, every subject—meet people. Exchange contacts. Build relationships."

A slight nod.

"Build your team."

"Don't just get through it," he said. "Conquer it." "Because in life, it's never just you."

Before the moment could settle—"Look at my scores!"

Sean's voice cut through the room.

He stepped forward, frustrated.

"How is any of this even legitimate? Every exam favors people who think like the person who wrote it!"

The room tensed.

Sean wasn't entirely wrong.

But he wasn't entirely right either.

The teacher didn't respond immediately.

And the moment passed. After School

Styler stood outside, near the edge of the road where a small patch of grass met the pavement.

He looked tired. Quiet.

The Morrison Mod 5 pulled up in its familiar hum. Harold sat inside.

Styler stepped in without a word.

The door sealed, and Harold switched the vehicle into

autonomous mode.

As the car merged into traffic, Styler began speaking—slowly at first.

About ideas.

Plans.

Possibilities.

Harold listened.

Then shook his head slightly.

"Don't think outside the box too much," he said. "You'll lose people."

A glance toward Styler.

"They're all still inside it."

"It's something to think about," he added. "If no one understands how you think… you won't connect with them." The car approached an automated drive-through—

Peck and Dish.

"For your chicken and mix."

A line of vehicles waited ahead, engines silent, systems active.

In the car in front of them, a group of teenagers shouted and laughed.

"Woooo! Party!"

"You robots don't know anything about that!" Harold placed an order.

A mixed drink.

Styler noticed immediately.

"I thought you quit."

There had once been a global prohibition—decades long.

A necessary shift during the transition from human-driven vehicles to fully automated systems.

It wasn't just about safety.

It was about discipline.

Productivity.

Survival.

Some even said it was part of a larger competition between planets.

Styler didn't like it.

Not at all.

But Harold—

Harold seemed to dislike himself for it even more.

"We both know alcohol should be illegal," Styler said quietly.

Then, without waiting, he turned on the radio. A small act of protest.

"Turn it off," Harold said.

Styler glanced at him.

"All music is unproductive," Harold continued.

"Doesn't matter the genre."

"It doesn't help anything. And it takes away from real connection."

Styler smiled slightly.

"Then… you want to talk?" A pause.

Then Harold nodded.

"Yes."

He kept his eyes forward as he spoke.

"When your mother… my daughter… passed…" His voice lowered.

"She asked me to take care of you." He reached into his pocket.

Took something out. Held it in his palm.

"A ring," he said. "It was hers."

"I've kept it with me all this time." A breath.

"Now… it's yours." Styler looked at it. Carefully.

Quietly.

He took the ring and placed it in his pocket. Gently.

And for the first time in a while—He didn't feel entirely alone.

6
FLOATING

Morning came clear and bright.

Sunlight streamed through the kitchen windows—translucent solar panels that absorbed light and quietly converted it into energy, feeding the house as the day began.

Styler stood at the center table, cooking.

The matte gray surface doubled as a heating element, warming evenly beneath the pan. He worked carefully—quietly—trying not to make a sound.

He flipped a pancake. Paused.

The first one, as always, wasn't right.

He set it aside.

The next ones came out even better.

Golden.

Controlled.

Once satisfied, he placed the finished stack onto a temperature-sensitive plate. It glowed faintly red at first—too hot—then slowly shifted to a soft pearl white.

Safe.

At the sink, he activated the food pipe. Fresh berries dispensed into his hand. He placed them beside the pancakes.

Simple.

Balanced.

He arranged everything onto a woven black serving tray—careful, deliberate.

Next to the plate, he placed something else.

A puzzle.

He carried the tray into his bedroom.

Quietly.

The room was sparse.

Not empty—but temporary.

The bed sat low to the ground.

A clear storage box served as a nightstand.

Clothes remained packed in suitcases instead of drawers.

Everything was ready to move.

At any time.

Kailee was still asleep.

Curled inside a sleeping bag on the floor, its built-in pillow tucked beneath her head.

The night before, she had insisted on taking the floor.

Styler hadn't argued—but he hadn't liked it either.

She stirred.

Slowly waking.

Styler stepped closer and gently set the tray across her lap.

She blinked, adjusting to the light.

He smiled.

"Put the pieces together," he said, almost playfully.

"You'll figure out what we're doing today." Kailee sat up, amused despite herself. "Alright," she said.

She began assembling the puzzle, taking small bites of the pancakes between pieces.

Piece by piece—The image formed.

Water.

A shoreline.

A kayak.

She looked up.

Understanding.

Styler watched her, then drifted briefly into thought.

Planets.

Gravity.

The balance between attraction and distance.

Like a relationship.

Strong enough to hold—Not so strong it breaks.

They spent the rest of the morning at the house.

Styler showed her his projects.

The vehicle.

The ideas he was building toward.

Outside, Harold worked.

Gloves.

Mask.

Face shield.

In his hands—a plasma arc welder, pulling gases directly from the air, eliminating the need for fuel.

It sparked with controlled intensity.

Powerful.

Unproven.

New inventions always carried a question: How do you measure safety—when nothing like it has existed before? Harold glanced over briefly.

Gave a small nod.

Then returned to his work.

By early afternoon, Styler and Kailee were ready.

They headed out around two.

Inside, Harold had finished for the day.

Now he sat, idly passing time with a simple single-player game—matching shapes, repeating patterns.

Styler stepped in.

"A single-player accomplishment," he said lightly.

"Who's even going to notice?"

"Can we take the auto?" Styler asked.

Harold didn't hesitate.

"Go ahead."

They took the Morrison Mod 5.

Its autonomous system made it legal for them to travel alone.

As they settled into their seats, the harnesses adjusted automatically.

The radio activated softly.

The car moved.

The city gave way to open land.

Miles of farmland stretched outward—orderly, efficient.

Automated systems moved through orchards and fields.

Apples.

Wheat.

Engineered crops.

Even sod farms.

They passed a hybrid grove.

Orapple.

A blend of apple and orange—created to end a debate no one could seem to settle.

Food no longer came from just one place.

Entire supply chains stretched across planets.

Some crops were grown light-years away.

Harvested.

Processed.

Shipped.

On one world, beneath a yellow sky, plants grew in shades of purple instead of green.

There were new foods.

New animals.

Deraal fruit.

Torquins—similar to cattle.

Licem grain.

Another planet—Ryhope—produced Furntiien, Hollis, and other unfamiliar resources.

All integrated.

All part of the same system.

Inside the car, Styler handed Kailee something small.

"A Mini Bake Cake."

She smiled and took it immediately.

Gone in moments.

The radio shifted.

A familiar song played—by Repay!. Kailee started singing along.

"In life is what it comes down to… I'm lucky I found you…"

Styler joined in.

She laughed softly.

The music softened.

The road stretched ahead.

Styler glanced at her.

"You don't get to drive much, do you?" She shook her head slightly.

He smiled.

"We can glitch it into manual mode." A pause.

"Want to drive?"

"I don't know how to drive," Kailee admitted.

Styler grinned.

"Want to learn?"

The Morrison Mod 5 eased itself to the side of the triway. The road stretched empty in both directions—no traffic, no interruptions.

Styler switched the vehicle into manual mode. "No restrictions," he said lightly. "Your turn." They swapped seats.

Styler guided her through the controls—acceleration, braking, steering.

"Driving's not really something you explain," he said. "It's something you feel."

A pause.

"The car will always be there—with or without you." Kailee placed her hands on the controls.

The car moved.

Smooth.

Natural.

Styler leaned back, watching her with quiet satisfaction.

"Maybe," he added, half-thinking aloud, "you always drive back to the person who taught you… or to where things began."

He smiled.

"So now—you'll always find your way back to me." Kailee didn't answer.

But she smiled.

The road opened wide.

Styler turned his gaze outward, idly counting passing vehicles, guessing how many people were inside—like a game.

When the cars disappeared, he counted trees. When the trees disappeared—there was only water.

Waves stretched across the horizon. In his mind, shapes formed—

A boat made of water.

A boat made of sand.

Kailee kept driving—steady, effortless.

Eventually, she exhaled.

"I'm tired," she said.

Styler nodded and returned the car to autonomous mode.

They arrived at the inlet. Waiting for them—Two kayaks.

Tethered to a floating dock.

Semi-translucent, just like in the puzzle.

Water shimmered beneath them, visible through their surfaces.

They pushed off.

Each in their own kayak.

Hours passed quietly.

The rhythm of paddling.

Water against the hull.

Silence, shared.

Dusk approached.

"Just a little longer," Styler said.

Then, pointing—

"Over there."

They rounded a bend.

And then—

Light.

Strings of warm lights stretched across trees along a hidden shoreline.

As darkness settled, they flickered on one by one—soft, glowing, alive.

Kailee stopped paddling.

"Styler… you did this?"

Her voice carried awe.

"I've never seen a real Christmas tree," she said.

"This is… the closest I've ever felt to one."

On the way back, the sky deepened.

Stars emerged.

Kailee looked up, eyes wide.

"Constellations matter," she said.

"If you know where they are—and where you are—you'll always find your way."

They returned to shore.

Styler activated his holocaster, summoning the Mod 5.

Moments later, it arrived—empty, waiting.

He opened the door for her.

Inside, the car hummed back to life.

Music played softly.

After a while, Styler lowered the volume.

"You know… we don't really listen to music anymore," he said.

"We replay the same songs until they lose meaning."

A pause.

"We miss the ones happening in real life." Kailee smiled.

"That's deep," she teased. "So you don't like music?"

"I do," Styler said.

"That's what scares me."

The car moved steadily along the road. Up ahead—

A man walking alone.

"We should give him a ride," Kailee said.

Styler hesitated.

"We're moving too fast—" Then reconsidered.

"Wait."

He switched to manual. Turned the car around.

Pulled up beside the man. "Need a ride?"

They dropped him off a few miles ahead.

A small stop.

Simple.

But meaningful.

Back Home

Styler's house was decorated—ribbons woven through the space in soft red and yellow strands.

Kailee stepped inside first. Followed them into his room.

On the bed—

A box.
Wrapped.
Waiting.
She opened it carefully.
Inside—
A balloon rose slowly, tethered by a thin string.
Attached to it—
A ring.
Floating.
Waiting.
Styler stepped closer.
"This is my commitment ring," he said.
A pause.
"It was my mother's." Kailee froze.
"I can't take that," she said softly. "It's too soon." Styler smiled.
"Not for me."
She took his hand instead, closing his fingers around the ring.
"Then hold onto it," she said. "For when I'm ready."
The Next Morning
Light filtered into the room.
Styler woke first.
Kailee still slept peacefully in his bed
Later, Harold drove them toward the city.
Along the way, they passed a new development.

A massive flying 3D printer hovered above empty land, constructing entire buildings layer by layer. Its spiral fans held it steady as it printed structures from raw materials—some gathered directly from the environment.

It was the same technology used to build on other planets. Cities. Factories. Entire colonies.
Then—
The car stopped.
A bear.
Then another.
Then more.
They surrounded the vehicle.

One climbed onto the hood.

Claws scraping metal.

Kailee woke abruptly. "What's happening?" Harold stayed calm.

"Stay still," he said. "Like the old films—say nothing." The bears pressed closer.

The car took damage.

Scratches.

Dents.

Then— Harold moved.

The car pushed forward, forcing its way through.

Moments later, a tow drone was already on its way.

The car would be repaired.

Everything could be repaired.

They dropped Kailee off at her building.

Later

Styler stayed behind in the city.

He had work to do.

Fog rolled in that morning—thick, low, controlled.

Systems built into the skyscrapers pulled moisture from the air, condensing it into usable water.

It streamed downward, collected, redistributed.

Even the air was engineered.

Styler entered a shop he had visited before.

This time, he noticed the manager.

A tall, heavyset man in a black uniform.

On the back—"Every Home, for everyone's home needs."

On the shelves—Glass candles.

Styler recognized them immediately.

Carbocandles.

Devices that burned carbon dioxide, purifying the air.

Used for atmosphere control.

Emergency use.

Environmental balance.

He stepped closer.

Interested.

Focused.

Styler's attention shifted to another display.

Carbosticks.

Slender, glowing rods that absorbed carbon dioxide and released oxygen. When someone breathed on them, they brightened—reacting instantly.

They were popular in clubs. Useful underwater.

Alive in a way most tools weren't. Styler tilted his head slightly, thinking. What if fireworks worked like this?

Something that didn't pollute—But reversed it.

He stepped closer to the counter.

The manager turned suddenly—Startled.

A glass candle slipped from his hands.

Shattered across the floor.

Light fractured in every direction.

Styler froze.

Then—Ran.

Out the door.

Onto the street.

He stopped a few steps away, breath catching.

Why did I just do that?

It wasn't fear exactly.

Not danger. Just—Instinct.

He exhaled.

Looked back at the shop.

Then turned around.

Went back inside.

The manager was already cleaning up.

Styler stepped forward.

"Can I pay for that?" he asked.

The manager looked up, surprised again—but differently this time.

Styler added, "I think I startled you." Meanwhile—Kailee's apartment the kitchen unfolded from the wall.

Table.

Seats.

All lowering smoothly into place with pneumatic precision—designed to maximize space.

Kailee sat across from her father.

Stephen.

There was a pause. Then—

"Are you sure he's good for you?" Stephen asked.

Kailee frowned slightly.

"Who?"

"Styler."

Kailee didn't hesitate.

"He's brilliant," she said. "He's… incredible." Stephen's tone shifted.

"He's from the outskirts." Kailee's expression hardened.

"We care about each other," she said. "That's enough."

A beat.

"Let's not do this." Back at the Shop

The manager finished sweeping up the glass.

Styler lingered nearby.

"How's your day going?" he asked.

The manager smiled faintly.

"It's going."

Styler glanced around.

"I come here a lot," he said. "It's one of those places I could work in… maybe even live in."

The manager laughed.

"Then you'd be stealing my house," he said. "My bed's in the back."

Styler smiled, then turned to leave.

As he stepped outside, voices caught his attention.

Raised.

Focused.

A negotiation.

A bartering agent stood beside a polished silver vehicle.

"I've had multiple offers," the agent said. "I'll need time to decide."

A pause.

"I want ten of the new Mods for it." No currency.

Just exchange.

Value for value.

Styler watched for a moment.

Then moved on.

The Auction House.

The building was older—ornate, deliberate in its design.

Inside, rows of seats faced a central stage.

Each seat had a simple bidding panel.

Some participants were present.

Others joined remotely.

Styler took a seat beside an older woman.

She glanced at him.

He smiled.

"You'd be perfect for my grandfather," he said casually.

She laughed.

"Oh? And you've known me how long?"

A small pause.

"If you knew me," she added, smiling, "you'd know I'd be perfect for any grandfather."

She studied him more closely.

"What about you?" Styler tilted his head.

"You've got that look," she said. "Like there's someone you don't want to lose."

A knowing smile.

"You're smitten."

She leaned back slightly.

"You know how these things go."

"Emotional beginnings… tend to have emotional endings."

Styler didn't look away. "Yeah," he said quietly. "I'm in love."

The woman nodded.

Softly.

"I've been in one of those before."

7

MY' LADY

Walking side by side through the city, Styler and Kailee moved between buildings that seemed impossibly tall—structures that stretched upward until they blurred into the sky. Robotic cleaning units climbed their surfaces, polishing glass and tracing every edge with mechanical precision. Along the streets below, smaller units moved quickly between pedestrians, keeping the ground spotless.

Eventually, they arrived at Your Pick Time Travel. An attendant greeted them with practiced efficiency.

"I've got you on the clock for three hours," he said. "That's sixty-three hour credits for the two of you."

Styler reached for his holocaster, but Kailee stepped in first and completed the payment.

"You didn't have to do that," Styler said.

Kailee smiled lightly. "I've got money. It's only fair.

Your family works too hard—I couldn't just let you pay."

The attendant led them down a corridor to two wardrobe rooms facing each other.

"You'll need to remove everything and change into the provided clothing," he instructed. "No modern items allowed."

Styler smirked. "Great. Old clothing." Inside the men's room, Styler paused.

The garments weren't worn or faded as he expected. They were immaculate—Victorian-era clothing, perfectly preserved, spotless, as if untouched by time itself.

He dressed quickly, unconcerned with appearance.

Who cares what I look like.

He placed his belongings—his clothes, his holocaster—into a cubby before stepping out.

Then he waited.

And for Styler, waiting always stretched longer than it should.

He slipped his hands into empty pockets, pulled them out, then repeated the motion—restless, passing time in small, meaningless gestures.

The door opened.

He looked up.

And froze.

Kailee stepped out in a golden dress—layered, flowing, detailed with delicate embroidery. White fabric framed the under layers, and the cut of the neckline gave the entire look an effortless elegance.

For a moment, Styler said nothing.

Then he smiled.

"Ready to go?"

Kailee lifted her chin slightly, already slipping into character.

"Have I ever been?"

They stepped forward together and passed through a clear scanning frame. It glowed briefly—

Green.

Cleared.

Ahead of them stood two portals.

Time waypoints.

One for departure.

One for arrival.

Both illuminated from below by a soft orange light.

The departure gateway opened.

And together, they stepped through.

For Styler, it was his first time.

And it wasn't what he expected.

There was no chaos.

No overwhelming sensation.

Instead—

Calm.

A strange, quiet stillness.

As if everything mattered deeply… and at the same time, nothing mattered at all.

Stress dissolved.

Conflict disappeared.

For a moment, there was only presence.

Then—

They arrived.

A crowd surrounded them.

Voices, movement, life.

Styler adjusted instinctively, then bowed slightly toward Kailee, his Victorian suit settling around him.

Above them, a grand chandelier shimmered with candlelight, its glow reflecting across the room. The light scattered softly, catching on Kailee's dress as she moved.

The walls were lined with oil paintings—rich, detailed, timeless.

A long oak table stretched across the room, covered with carefully arranged hors d'oeuvres.

Kailee looked around, visibly amazed.

"How did you know I would like this?" Styler smiled, a quiet confidence in his voice. "From the book you were reading at Fill."

She didn't answer immediately.

Instead, she turned and drifted slowly into the room, her sun-dried yellow dress flowing behind her as she took everything in.

Kailee glanced back at Styler, her eyes bright.

"I really like this time period."

She tilted her head slightly. "Good intuition.

Can you read me now?"

Styler smiled, energized by the moment.

"Let's dance."

For a brief second, something flickered—like time itself hesitated—then the music carried them.

They moved together across the polished floor.

Kailee smiled. "Are you enjoying yourself too?" Styler leaned closer, his voice low.

"As long as I'm with you, I wouldn't care where I am… or what's around me."

Around them, the room shifted into celebration.

Glasses were raised.

"To a virtuous life!" shouted a man in a rumpled waistcoat.

The crowd echoed him—Styler and Kailee included.

"To a virtuous life!" They drank.

And then—Something changed.

The man slipped away.

A woman staggered.

Another guest dropped to their knees.

Within seconds, people began collapsing—some vomiting, others trembling, some going rigid and still.

The music faltered.

A voice cut through the chaos.

"Only others can notice when you die," a man said coldly. "You never notice it yourself."

A scream broke out—a young woman clutching her fiancé's lifeless body.

Nearby, another man whispered over a corpse, "Sleep the eternal sleep."

Kailee froze.

Styler pulled her close, resting her head against his shoulder.

"It's going to be okay," he said quietly.

But the voices continued.

"Were you ever owed a life?" someone murmured from the corner. "Did you earn it?"

Another voice, louder now—

"You have been granted a slow, painful death." "Speakers unite!"

Styler's eyes darted across the room.

Something wasn't right.

Nearby, a vase held fresh plants.

He moved quickly, pulling two stems together, twisting them into a crude test.

If the drink was poisoned, the plant would react—wilt, discolor, die.

He dipped it into one of the glasses.

Nothing.

No change.

Styler whispered under his breath, "Don't give up on me…"

Then—

Laughter.

Bodies rose.

Guests straightened.

The dead were not dead.

They were smiling.

The entire room shifted again—alive, theatrical.

A performance.

A deception.

A test.

Styler exhaled sharply.

"So this is a Speakers of Nadjin gathering…" he said. "What kind of test is this?"

Kailee shook her head, unsettled.

"I don't know… but we failed it."

Moments later, they stepped back through the portal into their own time.

Inside Your Pick Time Travel, Styler turned to the attendant, visibly shaken.

"What kind of horror story did you send us into?"

The attendant studied his expression, barely holding back a laugh.

"You want a refund?"

"You traumatized me," Styler shot back. "Yes, I want a refund."

Kailee sighed, stepping in gently.

"Oh, come on, Styler." "Let's just go."

They parted soon after, heading in different directions.

The Next Day

The class gathered outside, walking toward a grassy clearing.

Sean bumped into Styler.

"Oh. You again."

Styler didn't react. "Did you do that on purpose? Just leave me alone."

The students sat in a loose circle on the grass.

Styler picked up a small dark stone, turning it over in his hands.

The teacher noticed.

"Sometimes the most valuable things are right in front of you," he said calmly. "You just don't recognize them."

He nodded toward the rock.

"Think about it—how much of our world, our technology, started with something as simple as this."

He smiled.

"Start collecting interesting ones.

They're far more useful than throwing them."

A pause.

"Would you like to be thrown away?" Styler muttered under his breath,

"Throw it at Sean… the rock wouldn't mind." Still, he slipped it into his pocket.

The teacher noticed—and said nothing.

Wind moved through the trees, scattering leaves.

Nearby, Jessica absentmindedly played with grass.

"Bored?" the teacher asked.

She shrugged. "Yeah." He nodded.

"In nature, what comes first usually exists in greater numbers. Grass came before us—and look how much of it there is compared to us."

Later, Dave leaned toward Styler, working on paper.

"I'm catching up on last week's assignment." "Why paper?" Styler asked. "Why not digital?" Dave shrugged.

"I like it. You draw your own lines. It builds precision. It's a blank canvas—you can do anything with it." Styler nodded slowly.

"I might try that."

The teacher raised his voice slightly.

"Jessica—what do you want to be?"

She smirked. "Same as my mom. Easy path. Guaranteed job."

The teacher's tone softened.

"Life doesn't guarantee anything. And you're not your parents."

He looked around at the class.

"You have to find your own direction—and carry that ability

with you."

The class settled.

The teacher concluded:

"Life isn't about the roads you're given." A pause.

"It's about the ones you create." "Good day."

Later — The City

Styler crossed into the city when a voice shouted—"Stylerrrrrr!"

Cody waved from across the street.

Styler and Kailee both turned as he approached.

"What are you two doing out here?" Cody asked casually.

Then, grinning—

"Stay out a little longer. There's a concert tonight. I'll get you in."

The Concert

The venue was packed—dark, electric, alive.

But the performers weren't human.

They were holonoids—perfectly rendered holographic performers.

A five-member girl group appeared, moving in flawless synchronization.

Holonoids were holograms capable of moving physical objects; in this case high tech guitars.

Their voices were artificial—But indistinguishable from real. Thousands watched.

Entranced.

Styler's thoughts drifted.

Why do these draw ten thousand people… but real performers don't?

What else could this technology do?

Furniture.

Structures.

Entire environments.

He barely noticed the performance ending.

Kailee, on the other hand, was glowing—completely immersed.

Afterward, walking together, Styler spoke absentmindedly:

"You know… hair turns white to retain moisture and regulate

heat. Like snow."

Kailee smiled.

"That's… oddly sweet."

But Styler's mind had shifted.

Toward Harold. Toward home.

Later, he typed quietly into his holocaster:

"I'm failing my grandfather. He's never failed me. I haven't been there enough."

Harold's Birthday

The next day, Styler went to visit him at work.

"I've always loved this place," Styler told the foreman. "I brought him something."

He revealed a floating chocolate ganache cake.

The foreman laughed.

"Our vending machines are already luxury." The sparklers lit themselves as Styler sang. "Happy birthday…"

The cake sang back.

Harold stood nearby, looking at old photographs—fading, fragile.

Styler stepped closer.

"Your past is supposed to stay behind you," he said. "So you can focus on the future."

Harold didn't look up.

"Your past," he replied quietly, "is part of your future."

8

KAILEE'S APARTMENT

Styler sat quietly aboard the automated transit system known as The Rail.

The train moved with a smooth, almost weightless glide—an alternating maglev that connected the outskirts to the city. From the outside, it appeared as a dark, matte surface, designed not to disrupt the night skyline. But the exterior wasn't fixed—it was a seamless, programmable display. It could shift into advertisements, camouflage, or ambient visuals depending on demand.

Inside, the carriages were just as engineered.

They were washed nightly at the depot—thoroughly. Grated flooring allowed for full pressure cleaning, a design that doubled as a drainage system in case of flooding or system failure.

Everything had a purpose.

Styler chose his seat carefully—near people who looked quiet, harmless.

He was on his way to Kailee's apartment for the first time.

Excitement sat in his chest—But so did something else.

A low, turning anxiety.

At some point, without realizing it—He fell asleep.

He was drifting.

Rising.

Floating beyond the atmosphere, into the vast black of space.

Hydrogen.

Helium.

Pressure.

Silence.

He couldn't breathe.

The dream shifted.

Humanity had reached the sun.

Not once—but repeatedly.

Learning.

Adapting.

Even… teaching.

Teaching the sun when to shine.

The moon when to dim.

Controlling day.

Controlling night.

Even the weather.

If animals can learn commands, he thought, Why not everything else?

Plants.

Atmosphere.

Storms.

Maybe the world itself could be taught.

Then—

Air vanished.

His chest tightened—Styler woke up sharply.

The dream dissolved almost instantly, slipping away before he could fully grasp it.

Outside the window, the city had come into view.

It didn't feel like something built.

It felt like something that had moved—like a mountain shifting into place just for him.

Arrival.

Kailee's apartment building rose among the city structures, sleek and precise.

As Styler approached, the door unlocked automatically—retina recognition through her holocaster system.

He paused for a moment.

The difference was immediate.

The outskirts felt distant now.

Outdated.

Forgotten.

Inside, the space opened into a level of refinement Styler wasn't used to.

On the wall near the entrance, a glowing installation pulsed softly—interlocking neon circles, layered with depth.

Styler pointed immediately.

"That's a Sunnie piece. I know his work." Kailee smiled.

"You do know your art.

My dad got it directly from him."

She stepped aside.

"Welcome. Want something to drink?" Styler hesitated.

"Water… or maybe Zox, if you have it." Zox—the shifting beverage.

Its colors moved like a living spectrum, changing with perception.

Some variants adapted flavor based on mood or pairing, turning every sip into something new.

The Apartment.

Styler's eyes moved slowly across the kitchen.

Everything was integrated.

Refined.

Efficient.

A single unit handled dishes and laundry—washing, drying, even folding.

It could heat, sanitize, and even cook.

The refrigerator was more like a system than an appliance—coffee, blended drinks, soda, all produced from dissolvable cartridges.

No waste.

Fully contained.

Transparent doors revealed everything inside.

Nothing hidden.

Shelving replaced cabinets—open, clean, intentional.

Every item faced outward, emphasizing function over appearance.

Nutrition labels visible.

Choices deliberate.

Styler felt it immediately.

Out of place.

"Do you have a bathroom I can use?" he asked, trying to steady himself.

Kailee nodded.

"Down the hall, to the right."

The Bathroom

Styler stepped inside—and stopped.

The shower system was unlike anything he had seen.

A single motion-responsive unit handled everything—water, soap, drying. Dissolvable cartridges fed directly into the system.

No clutter.

No effort.

He moved closer, almost unconsciously.

Like a moth drawn to light.

The mirror didn't fog. Didn't obstruct.

And when he reached toward it—His hand passed through.

Accessing stored items behind it.

He barely noticed how long he had been there. A knock.

"Hey," Kailee called from outside. "What's taking so long?"

A pause.

Then teasing—

"What are you selling in there? Looks aren't a product—you should know that." Styler snapped out of it.

He opened the door—

And nearly collided with her.

For a brief second, they were too close.

Close enough to kiss.

Both paused.

Just for a moment.

Then the moment passed.

Styler stepped out, quickly rinsing his hands again—watching as the system dried them automatically.

No towels.

No effort.

He glanced back once more at the faucet.

Still slightly in awe.

Styler stood there, quiet—lost in his own thoughts.

It wasn't unusual for him.

If anything, it was where he lived most of the time. Kailee watched him for a moment, then smiled. "Did you finish the broken puzzle?"

He blinked, pulled back slightly.

"My dad always asks that," she added. "Whenever someone looks like they're thinking too hard."

Styler let out a small breath, still taking everything in.

"Some of the technology you have here… it's incredible."

Kailee shrugged lightly, unsure how to respond.

"If that impressed you," she said, "there's a pool in this building where the floor rises from the bottom. It even lights up as you step on it."

A small pause.

"We could check it out." Styler shook his head gently. "Maybe another time."

Kailee reached up and pulled down a reclining panel built into the wall. Matching chairs unfolded with it, settling into place smoothly.

They both sat.

She prepared drinks from the refrigerator system without effort, handing one to him.

Time passed quietly.

Then Styler glanced at the clock.

"Look at the time… I should probably go." He stood, moving toward the door.

But just as he stepped out—Kailee's parents arrived.

Later — Home

Styler burst in, already mid-thought.

"You should see the shower Kailee has! It makes you wonder what towels are even for."

Harold looked up, deadpan. "Reusable toilet paper." Styler laughed.

"Maybe. You washing them?" Harold didn't smile.

"The key to a joke," he said calmly, "is delivering it with a straight face."

A beat.

"You craft the experience." Styler smirked.

"That… was a joke. I'll take it."

But Harold didn't respond right away.

Something about him felt different.

Quieter.

Distant.

"There's more I want to talk to you about," he said.

His gaze lingered—heavier than usual.

Styler shifted slightly.

"Grandfather… I don't think that's a joke." Harold exhaled slowly.

"You know… that's a lot to live up to."

He moved through the room as he spoke, his voice steady but reflective.

"Some of the things we spend time on in life…" "…they're a waste of it."

A pause.

"We only get so much time." He looked at Styler.

"You're young. But at any age—you have to find yourself."

Harold's attention drifted toward the vehicle they had been building together.

He picked up a carbon fiber piece, examining it.

"You never really feel the accomplishment of solving something no one else has solved," he said. "Not truly."

"When it's recognized… the feeling comes from others. From them seeing it."

He set the piece down, smiling faintly.

"I'm not even sure I did this part right." He turned back to Styler.

"There's something else you'll learn."

"You can do a hundred things right… and one thing wrong…"

"And that's what people remember." Styler hesitated, then spoke.

"I was at Kailee's place today." A pause.

"What if I'm not good enough for her?" Harold studied him.

"Right now, Styler… we're not in sync." "Our minds are in different places."

He stepped closer.

"My life is coming back to me," he said quietly. "And yours is just beginning."

A beat.

"Not everything is about you."

"You're not the main character in a story." Silence settled between them.

9

A CHRISTMAS

A bright white layer of snow stretched across everything—rooftops, roads, fields—blanketing the world in silence.

Miles of it.

Unbroken.

Styler stirred awake in bed.

Still half-asleep, he slipped on his holocaster and pulled up the morning forecast.

"Sandra, what have you got for us today?" The broadcast flickered to life.

"Ohhh, it's going to be a white Christmas!" Sandra said cheerfully. "Snow, snow, snow. Santa's touching down softly tonight."

She smiled.

"And in case I'm too full of eggnog to make it in tomorrow—Merry Christmas, everyone."

Morning.

Styler rushed through his routine.

He brushed his teeth in the shower—saving time.

Toothbrushes weren't what they used to be.

No bristles, no paste.

Just a slim light-wand that cleaned and refreshed instantly.

Shaving tools looked nearly identical.

Easy to mix up.

Outside, the cold deepened.

Inside, it was quiet.

Styler and Harold sat across from each other at the round kitchen table.

Christmas Eve morning.

Just the two of them.

A synthetic tree stood in the corner—lit, decorated, but understated.

Neither of them seemed to care whether it was there or not.

They ate in silence.

Styler stared at his food—the way oil separated and pooled across the surface.

And his mind drifted.

Everything we see as whole… is made of parts.

Lipids. Molecules.

Systems building into systems.

He could almost imagine an entire class built around the thought.

Then, quietly—

"Where do I fit in the world?" Harold didn't miss a beat.

He reached under the table, pulling out a wrapped package.

"In about a medium." Styler blinked—Then laughed.

He stood and hugged him. "Merry Christmas." "Merry Christmas, Styler."

He unwrapped the gift and pulled on the new shirt. It fit.

Of course it did.

They sat again.

Quiet.

Reflection.

Christmas had always confused Styler.

The lists.

The wanting.

The asking.

As a child, he never understood why people asked for things beyond what they needed.

Food.

Warmth.

Water.

Where did toys come from?

Why did children want objects no one in their family had made?

It always felt… taught.

Like the desire itself had been placed there.

Are people taught to be high-maintenance? He wondered.

Christmas Day. Snow fell heavier.
Decorations filled the streets.
And yet—the roads and rails remained perfectly clear.
Styler arrived outside Kailee's apartment building.
He waited.
The door opened.
Kailee stood there—smiling, a little mischievous.
"We have a surprise for you."
Inside the apartment had transformed.
Garlands, wreaths, soft lighting—everything carried warmth.
At the center stood a large synthetic Christmas tree.
But this one was different.
The ornaments told a story.
Animals—cow, chicken, lamb. Plants—apples, oranges, grains.
A full ecosystem represented in miniature.
A reminder:
Life feeds life.
There were also objects from history.
Ships.
Trains.
Planes.
Computers.
Even things as simple as air conditioners.
Each one a symbol.
Of progress.
Of survival.
Of the lives it took to build them.
Styler stood there, taking it in.
For a moment—
He imagined himself as part of it.
Part of a family.
The Kitchen
He wandered in quietly.
Kailee's mother stood there—mid-task.
Warm presence.

Calm energy.

In her hands was a device Styler rarely saw.

A Shopper.

A flat, rectangular interface.

You selected an item—

And reached inside to retrieve it.

As if space itself folded.

Styler watched carefully.

That shouldn't exist so casually, he thought.

"They're out," Kailee's mother said, frustrated.

"Completely out."

She looked at them.

"The Shopper doesn't have the essentials today." A small sigh.

"Why don't you two go to the store? The old-fashioned way."

Styler nodded immediately.

"No problem."

Kailee smiled.

"Yeah, Mom. We've got it."

And just like that—

They stepped back into the cold,

Heading out together to finish preparing Christmas the only way left—

By hand.

They walked side by side through the city, their breath faint in the cold air.

"It's colder today," Kailee said, pulling her coat closer. "When it gets like this, the days feel shorter too."

A family passed them—bundled tightly, layered against the cold.

Most of them had green hair, the now-common trait of their generation. In the distance, small flashes of red and pale blonde broke the pattern.

The Store.

The grocery store stood ahead, framed by faux garlands and synthetic wreaths.

Inside, it was unmistakably modern.

UV lighting washed over the packaged goods, revealing whether items had been handled or tampered with. Shelves were clean, minimal—products designed with embossed textures instead of heavy ink, reducing waste and making recycling easier.

Everything felt intentional.

The fresh section was different.

Alive.

At the press of a button, plants could grow instantly.

Animals too.

A demonstration unit activated—An embryo formed.

Then a calf.

Then—

A full-grown cow in seconds. And then—

Failure.

The system glitched.

The structure destabilized, and the organism broke apart unnaturally.

Before anyone could react, sanitation robots swept in, clearing everything with clinical precision.

The system reset.

As if nothing had happened.

Styler stared.

Not amazed this time.

Just… unsettled.

Nearby, the florist section floated.

Individual flowers hovered in place, suspended by controlled air density and subtle airflow systems.

No stems, no strings—just motionless blooms hanging in space.

Kailee gently nudged him forward.

"Come on."

They gathered what her mother needed.

Paid.

And stepped back outside.

The snow had thickened.

What had been steady was now heavy—falling faster, covering the city in a deeper white.

Back at the Apartment.

Warmth met them immediately.

Soft.

Comforting.

Alive.

Kailee reached for Styler's hand and pulled him gently toward the dining room.

It was the first time they held hands like that.

Fully.

Knowingly.

The Table.

The table was large.

Rectangular.

Covered in deep red velvet.

Styler noticed instantly.

This isn't neutral.

Rectangular tables meant structure.

Authority.

Hierarchy.

He already knew how it would unfold.

Her parents at either end.

The conversation shaped from there.

At the center sat a fluff cake.

Soft, layered, filled with pockets of different flavors—a tradition from the energy revolution era.

Kailee smiled at it.

"We make this every year. My mom's version is the best."

Above them, light pods projected soft holiday imagery

across the ceiling—trees, wreaths, shifting patterns of light. Kailee's mother, Samantha, noticed Styler watching. "You wouldn't be admiring my fluff cake, would you?" Styler straightened slightly.

"It looks incredible. I'm sure I'll enjoy every bite." Kailee gestured nearby.

"This is my younger sister, Constance." Styler nodded politely.

She looked about eight—bright, observant. Blonde hair that

might darken with age, like many did.

Constance carried a plant into the room, placing it carefully as decoration.

"You know," Kailee said lightly, "plants are smarter than people think. They don't move—but everything they need comes to them."

Styler smiled.

"So does my food. Delivery." Constance looked at him, half-offended. "You just praised a plant?"

Before it could turn into something more, Samantha stepped in warmly.

"We exchange gifts every Christmas evening." She turned to Styler.

"I'd like to give you mine first."

She handed him a neatly wrapped package.

Inside—

A hand-knitted gray scarf, intricately woven.

Styler paused.

Then smiled, genuinely.

"Thank you. I'll cherish this. It's… really beautiful." Stephen stepped forward next.

More reserved.

He handed over a second gift—perfectly wrapped, layered with precision.

Styler opened it slowly.

Inside—

A holocaster.

But not just any.

The Icon A4.

Sleek.

Minimal.

Designed to sit flush against the skin.

No visible interface.

No display.

This one didn't show you information.

It became it.

Signals transmitted directly into the brain—visuals, sound, even scent.

Reality layered seamlessly over perception.

You didn't look at it.

You experienced through it.

It could adjust voice, language, tone.

Even influence thought patterns.

Expand perception beyond natural limits.

Styler stared at it.

Overwhelmed.

Then, without thinking—"How can I accept this?"

Stephen leaned back slightly, studying Styler.

"Kailee talks a lot about you—your interest in science, inventing."

He gestured toward the gift. "I thought being on the cutting edge might suit you."

A faint, almost restrained smile crossed his face.

"These aren't even available in stores yet."

Before Styler could respond, constance rushed forward and wrapped her arms around him.

It caught him off guard.

A tear slipped down his cheek—quiet, unannounced.

For a moment, he didn't wipe it away.

She handed him her gift.

Inside, a transparent, indestructible mug, printed with the periodic elements.

They finished exchanging gifts.

It was late now.

Too late for Styler to reasonably make the trip home.

Samantha spoke gently.

"You don't have to leave tonight.

It's late—and it's a holiday."

Stephen added, more matter-of-fact: "You can take the spare room." Styler glanced around.

Their apartment alone was larger than his entire home.

There wasn't really a decision to make.

The Spare Room.

Alone, Styler sat on the edge of the bed, turning the Icon A4 over in his hands.

He hadn't expected anything this Christmas.

Certainly not this.

And yet—

Here he was.

He broke.

Quietly.

Tears came again, heavier now.

And somewhere in his mind, Harold's voice surfaced—steady, familiar:

Crying is natural.

It's part of the equation.

You learn how to solve it—but it still returns when it wants to.

Styler thought:

Why don't I have this?

If my parents were alive… would it feel like this?

The door opened softly.

Kailee stepped in.

"It's okay to cry," she said.

Styler shook his head slightly. "I don't want to talk about it." Kailee moved closer.

"You'll run into problems if you can't say things—even to your own family."

He looked up, sharper now. "What if you don't have one?" A pause.

Then—

"You can say anything to me." Styler exhaled, frustrated.

"Ugh…"

Kailee hesitated, then asked anyway:

"Why don't you just use a holocaster… to talk to your parents?"

Styler's voice dropped. "Because they're dead." A beat.

"I could… but I won't."

"They're not coming back. There's no past with them raising me. That part just… doesn't exist."

His eyes were wet again, but he barely reacted.

"I still feel it though."

A quiet, almost defeated breath. "I don't want to fix it anymore." They sat side by side.

Then—

Styler leaned over, resting his head in her lap.

Slowly, he drifted to sleep.

Morning.

The next morning, Styler tested the A4. Harold appeared—clear, immediate.

"How was work?" Styler asked.

"I made it back," Harold replied. Then, after a moment: "I've noticed something. You've been spending a lot of time with that girl."

Styler didn't answer.

Harold continued, calm but deliberate:

"I want you to be happy. I won't interfere." A pause.

"But think about this… The things that matter most are the ones that outlive us."

He looked directly at him.

"Is that what you have with her?" Styler ended the call.

The Stadium.

They sat at the breakfast table—Stephen had tickets.

A surprise.

"I don't want to go," Kailee muttered. Stephen didn't entertain it.

"If you stay inside all the time, you might as well be homeless."

Then, more firmly:

"We're going."

The stadium was massive.

Crowded.

Alive with motion.

Stephen raised his voice theatrically: "Welcome to the world of Holosport!"

"It's the HellCats versus the Nightmares!" The game unfolded—fast, controlled chaos.

A holographic ball split mid-play, duplicating across the field.

Players moved through projected armor, striking goals that floated in shifting space.

First to fifty wins.

Stephen leaned toward Styler.

"What do you think?" Styler shrugged slightly.

"It's interesting. I actually met someone whose father owns the HellCats."

Stephen scoffed.

"I think they're idiots."

He gestured toward the field.

"They risk injury for this? Risk losing the ability to build a life? A family?"

He leaned forward.

"Technology and science—that's what matters.

Creating something that lasts beyond you." He continued, more personal now:

"My family built homes. Real estate. That's what matters in a growing world."

"The only thing worth risking yourself for is the people you care about."

Styler went quiet.

A memory surfaced—Harold.

A younger Styler.

Sketching houses on scraps of paper.

"Before anything else—draw five houses. Solve those problems first. Anything else? That's just… paid homelessness."

Back in the present, Styler realized something.

Harold and Stephen—Different men.

Same philosophy.

"HellCats 36! Nightmares 22!" the announcer boomed.

A player slammed equipment in frustration. Penalty issued.

Kailee shrugged.

"They're rich. But money isn't everything." Stephen turned sharply.

"Money isn't everything?" He leaned closer.

"With control of currency—you control everything. Industry. War. Innovation."

"You can build—or destroy—entire worlds." Styler spoke quietly.

"What matters to me… is how people treat you when money isn't involved."

Kailee nudged him. "Can we leave?" Styler smiled slightly.

"Already thinking that." They said their goodbyes. Stephen nodded.

"We already know who's winning anyway." The City Again.

They wandered into a multi-level outdoor mall—spiraling upward, open to the air.

Heavy goods below. Clothing higher up. Electronics near the top.

Everything layered like the city itself. Around them—People in adaptive clothing.

Shoes that reshaped with each step.

Garments printed in real time.

Styler glanced at Kailee.

"You're close to home, right? I can walk you." She nodded.

As they walked, he spoke more seriously: "These past few days… meant a lot."

A pause.

"But I need to get back. To Harold." "He doesn't have anyone else."

Back Home.

Styler returned.

Harold.

Fire.

Familiar ground.

Later, alone, he explored the A4.

The experience was overwhelming—signals, sensations, layered reality.

Better than hallucinations, he thought.

He activated head-share with Kailee. Their thoughts

intertwined.

Music. Fragments. Ideas.

She distorted his playlist intentionally.

Playful.

He almost smiled. After the Snow Days passed.

The snow melted into rain.

Then more rain.

Three days straight.

Finally—

"The Screen" activated.

A massive projection system stretched across the sky.

Light formed structure.

Particles aligned. Rain stopped mid-air.

Collected.

Redirected.

The city stayed dry.

Controlled weather.

Engineered environment.

Efficient.

Impressive.

But not without risk.

And Styler knew—

Technology like this always came with a cost.

10
DREAM

Morning came early.

Harold had already been awake for some time—he knew Styler had plans today, and he made sure to be ready before him.

He prepared a small plate of fresh fruit and walked quietly toward Styler's room.

Styler was already up.

Dressed.

Leaning back against the wall, holocaster active, scrolling through detailed 3D renderings of consumer electronics—rotating models, cross-sections, internal layouts.

He wasn't browsing.

He was studying.

Thinking.

How could any of this be applied to the vehicle? Harold stepped in.

"I brought you some breakfast." He set the plate down.

"Eat up. Then I'll take you into the city—to meet Kailee."

A short while later—Styler hurried outside, finishing the last of the fruit as he moved.

Harold was already by the Morrison Mod 5, wiping down the exterior.

The car hummed quietly, idling, warm. Harold noticed him.

"Hop in."

Styler opened the door carefully and slid into the passenger seat.

They pulled out of the neighborhood.

Something felt… off.

Houses lined the street with sale markers.

Too many.

Harold exhaled, shaking his head.

"A monument still belongs to its makers. A statue doesn't change. A palace belongs to royalty."

A pause.

"Then why does real estate keep changing hands?" He glanced out the window again.

"Since mass production, some people have lived generously without ever earning it."

His tone softened slightly.

"Styler… be grateful for what you have. Always." They passed another house.

Then another.

Harold gestured subtly.

"Over there—those people… always looking to take.

Never to build."

He scoffed lightly.

"I don't even know how they kept their home.

Probably sympathy from people who actually work." A moment of silence.

Then—

"Not everyone thinks like you." He looked ahead, steady.

"The ideas of a few built everything around us." A beat.

"Just don't become someone who thinks they should be paid just to talk."

In the City.

Later, Styler and Kailee walked side by side through the

city.

Traffic flowed around them—autonomous, efficient.

One vehicle passed, completely wrapped in dynamic screens—ads shifting across every surface.

Another taxi.

Another spectacle.

Styler broke the silence, thinking aloud.

"In computer science, 3D space is mapped in X, Y, Z coordinates."

He gestured lightly.

"So if everything began moving in one direction… which one was it?"

Kailee glanced at him. Already bracing.

"Living things grow upward," Styler continued. "Height over width—like plants, like people."

"That's like a Y direction." "Water evaporates upward too." He shifted slightly.

"But we think left to right. Writing, reading—that's X." "Non-living systems shaped how we interpret space." He smiled faintly.

"I'll make it harder—what coordinate did existence start in?"

Kailee groaned.

"That's a brain thumper." She thought for a moment.

"Maybe X came first. Like… chromosomes? Females are XX—before males with XY?"

She looked back at him. "So what do you think?" Styler shrugged, amused.

"I don't know. That's why I asked." They both laughed softly.

Their walk carried them into Time Travel Avenue.

The atmosphere shifted instantly.

Different eras lined the street—each storefront pulling at history in its own exaggerated way.

They passed Medieval Time Travel, its exterior built like a stone fortress—towers, banners, artificial age.

Styler bowed dramatically.

"Mi'lady, is this our destination for the evening?" Kailee smiled, playing along, dipping into a curtsey. "No, my lordship. Just a little further."

She took his hand and led him onward.

They stopped at Discount Time Travel.

Styler eyed it.

"Hopefully 'discount' doesn't mean we get ripped apart." Kailee grinned.

"I picked this one for you." "Alright," he said simply.

Inside Discount Time Travel.

They split off into separate changing rooms.

Despite the name, the interior was surprisingly refined.

Pleated leather seating.

Ornate paneling.

A large intelligent mirror stood before Styler.

It scanned him.

Then displayed instructions:

Recommended attire.

Environmental wear.

Behavioral alignment.

He followed along casually.

From the other room—

"I'm ready!" Styler called out. "Just a second!" Kailee replied.

Inside her room, the mirror had already selected an outfit.

A conveyor system slid it forward—precise, deliberate.

Moments later, they stepped out.

Styler looked at her.

Then at himself.

He raised an eyebrow.

"Are we… supposed to look like dirtied workers?" Kailee frowned slightly.

"I just did what the mirror told me." Styler shrugged, grinning.

"Let's go anyway." A beat.

"I'm excited."

Arrival: The Early Industrial Age.

They emerged into a world choked by progress.

Only one nation had industrialized—fully, aggressively—and the rest of the world seemed to orbit its output.

Textile mills stretched across the horizon.

Coal smoke bled into the sky.

Rivers ran thick with waste.

Innovation was everywhere.

So was consequence.

Styler and Kailee moved through the streets, their clothing already soiled to match the time.

Styler wore a tattered dark grey suit.

Kailee's layered dress mirrored it—faded, worn, practical.

Styler's eyes locked onto a factory in the distance.

He pointed.

"Let's go there."

They hurried down the muddy street.

The Machine Shop

Inside, the air was dense—metal, oil, heat.

A steam engirre powered the entire operation.

Machines clanked in rhythm.

People worked in tight coordination.

"Hold!" a man shouted.

A press slammed down—metal screaming under force.

Styler watched closely.

He didn't need a blueprint.

He understood.

"They're making parts for an omnibus," he said under his breath.

In the distance, a train roared—one of the earliest of its kind.

Crude.

Powerful.

Styler imagined it against the systems of 2137—the seamless rail networks, the underwater freight lines.

This was the beginning of all of it. The Textile Mill

They moved into a connected factory.

The looms were everywhere.

Thousands of them.

Moving endlessly.

Thread weaving into fabric.

Machines weaving into danger.

Styler and Kailee slipped between them—almost dancing through the narrow paths.

It was mesmerizing.

And terrifying.

Workers—men, women, children—labored in worn clothing to produce something new.

Then—

Styler noticed her.

A young woman.

Dark curls.

One arm missing.

Still working.

Kailee froze.

The realization hit her harder than the noise.

Outside.

Kailee drifted away without a word.

Styler turned—

She was gone.

He stepped outside.

The street opened into a polluted stream—dyes swirling across its surface in unnatural colors.

Metal scraps.

Chemical runoff.

Kailee stood at the edge, staring into it.

Her reflection fractured by color.

Styler joined her.

He glanced down at his own reflection—And flinched.

"I scare myself sometimes," he said quietly.

Kailee didn't look away. "Stay in the real world." Styler turned to her.

"Which one is that? This… or 2137?" She finally met his eyes.

"The real world is wherever we are." Rising Tension.

A commotion grew.

A street performance nearby—ragged voices, strained rhythm:

"We old kings… never made it that far…" The song twisted into something darker: "We wrought iron… we wrought lead… We built machines to do it instead…"

Not polished.

Not perfect.

But real.

Then—Shouting.

A constable pushed through the crowd.

Two men fought in the street—drunk, unstable.

And beyond that—

A march.

A protest.

The Riot.

It escalated fast.

Too fast.

Kailee grabbed Styler's hand. "Let's go!"

The crowd surged.

A constable was knocked down.

People screamed.

Bodies pressed in from every direction.

Some fell.

Some didn't get back up.

Styler could barely hear himself think—only fragments.

Whispers of thought swallowed by noise.

Kailee pulled him forward.

He didn't hesitate.

He trusted her.

They forced their way through the chaos—Until the crowd thinned.

Until movement became possible again.

Aftermath.

They slowed.

Breathing hard.

Styler laughed—half exhilaration, half disbelief. "That was… incredible. Scary—but incredible." He looked at her.

"No one could've handled that better than you." They walked.

The tension still lingering.

Then Styler spoke again—quieter now.

"I've always wondered… how did you know where to find that book?"

Kailee smiled.

"I'd read it before." A small pause.

"Styler… I'm Arlen too." He stopped.

"I'm Arlenish," she continued. "I spent time learning about our culture."

She looked at him—soft, certain.

"I knew exactly where the book was." "And I knew you'd need it." Return… or not.

They reached the waypoint.
Stepped through.
Light folded around them.
And then—
They stopped.
Something was wrong.
They weren't where they were supposed to be.

11
ABANDONED

They stepped through the waypoint—

—and assumed they were home.

The Wrong Return.

The suburbs looked familiar at first.

Same roads.

Same layout.

Same quiet.

But something felt… abandoned.

As they approached Styler's house, the illusion broke.

Overgrowth swallowed the structure.

Vines climbed over windows.

Grass pushed through the foundation.

It wasn't just neglected.

It had been left.

Styler opened the door.

The house powered on automatically—systems still functioning.

Then the clock spoke.

Time.

Date.

Year.

2188.

They froze.

Looked at each other.

Neither spoke for a moment.

Then Styler, almost reflexively: "Hungry?"

Holding onto normal.

He moved toward the kitchen.

"The food pipe still works."

He placed containers beneath it and activated the system.

Protein pellets poured out—unchanged, preserved.

"These things last forever," he muttered.

He handed one to Kailee.

"I know it's not city food… not Louis Riley level… but it'll do."

She took a bite.

Paused.

Then continued eating.

Not bad.

For Styler, it felt familiar.

Grounding.

But something was wrong.

He could feel it.

Couldn't name it.

Waiting.

They stayed.

Waited.

Hours passed.

Styler wandered through the house—touching unfinished projects.

The vehicle.

Fragments of ideas he and Harold never completed.

He sat down.

A few pellets in hand.

"I don't know what to do anymore." A quiet breath.

"If Harold were here… he would."

Kailee watched him. Then suddenly—

"I've got it." She stood.

"What if we go back to Time Travel Avenue?" "We just… leave this year."

Styler looked up.

Then smiled.

"Brilliant."

The Garage.

They moved quickly.

The garage was dim.

Dust covered everything.

Time had settled here.

Styler opened the door fully—And saw it.

The Morrison Mod 5.

Covered in soot.

But intact.

"We've got a ride!" The Drive

They took the triway.

The world had changed.

The Farsers—once active—were now buried under the very growth they were meant to control.

Machines overtaken by time.

After an hour, they reached the city.

Or what was left of it.

Ruins.

Time Travel Avenue was gone.

Collapsed.

Broken.

Silent.

A girl stood nearby—crying.

Throwing rocks.

Kailee approached gently.

"What's wrong?"

The girl shouted through tears:

"I can't get back!"

Styler felt it immediately.

They weren't alone.

"Name?" Kailee asked softly. "Shannon…"

She struggled to breathe.

"I used Luxury T Travel… and now I'm here." Kailee held her.

Styler stepped closer.

"Stick with us."

"What year are you from?"

"2137."

Styler's stomach dropped.

Kailee brightened.

"We're from the same year!" Styler shook his head.

"It's 2188."

His voice hardened.
“Our families… they’re gone.” “Everyone is gone.”
Kailee didn’t accept it. “We’re here.”
She looked between them. “We’re here together.” Stranded.
They returned to the car.
It wouldn’t start.
Once.
Twice.
Three times.
Nothing.
Even manual override failed.
Styler exhaled.
“I guess… we walk.”
The Entrance.
After miles on foot, they found it.
A massive concrete opening.
Half-open doors.
Leading underground.
A girl stood outside—watching.
Shannon approached first.
“Hi! What’s your name?” “Melissa,” she replied.
Then, simply:
“Want to come in?” Underground.
They followed.
The ground shifted underfoot—gravel, unstable.
Dim light flickered overhead.
Electricity still existed here.
Barely.
Inside—People.
About twenty-five.
Adults.
Children.
Survivors.
A fire burned faintly to one side.
Shannon drifted toward a boy her age—already trying to

normalize things.

Smiling.

Flirting.

Kailee leaned closer to Styler.

"What do we do?" Styler didn't hesitate. "For now?"

A breath.

"We survive."

Then, quieter—

"Be where you are."

"You choose how to live based on where you are." He looked at her.

"The problem is… we're always living in a future we don't even know exists."

"You have to live now."

Kailee nodded, but her voice trembled.

"I'm scared of my future."

"I don't know if everything will be okay." Styler gave a small, crooked smile.

"You're already in the future." The Children.

A child cried in the corner. Styler walked over.

Used his sleeve to wipe the tears.

"Hey…"

Softly.

"We just keep going." "Right foot. Left foot." Nearby, Shannon muttered:

"Look what people did to you…"

The room shifted uneasily. The Voice.

Someone covered Styler's eyes from behind. "Guess who?"

For a moment—He thought it was Kailee.

It wasn't.

Shannon laughed.

Then turned to the group.

"Despite everything… thank you." "You're not the family you had." "You're the family you make."

A fragile idea.

But it landed.

The Shift

A child began reciting:

"A B C D E F—"

Styler turned sharply.

"Quiet."

"We need to hear."

And then—

A voice.

From the dark. Low.

Measured.

Speaking Nadjin.

"-sen… -cha… -je… -den…" The man stepped forward slowly. Into the dim light.

"-ie… -ko… -ra… -chu… -en… -za…" He smiled.

"We're all going to die."

Silence fell.

Styler stepped forward.

"I refuse that."

He moved into the shadows. Meeting the man where he stood. Fear didn't push him back.

It pulled him closer. Because sometimes—What scares you most…

Is what you need to understand.

Styler looked at him steadily. "You speak Nadjin."

A pause.

"We're going to need you." They held each other's gaze.

Understanding passed between them unspoken, but complete.

The shrouded man smirked faintly.

"The mind can only take so much trauma," he said. "Are you ready for what you're stepping into?"

He tilted his head.

"You think you can fight them? We've been trying for nearly a decade."

Styler didn't flinch. "Who is we?"

The man shifted slightly, the fabric of his cloak brushing

the ground.

"I had a squadron.

Left it after the city fell… during the Great Battle of 2181."

Styler's voice sharpened. "What year is it?"

"2188," the man replied casually.

The number hit harder this time.

More than fifty years displaced.

The man studied him.

"Don't let your emotions take control," he added.

"If everyone did that, nothing would ever get done." The Fractured World.

Around them, the group tried to hold onto fragments of normalcy.

A few laughed.

Some played.

Most watched.

"I can teach you some of the language," the man offered.

Styler shook his head.

"I don't want to speak Nadjin."

A beat.

"I don't believe in learning other languages." The man chuckled darkly.

"Good. Because when someone commits a crime in a language you don't understand…"

He made a small motion across his throat.

"…you solve it quickly." Styler laughed despite himself.

"This is exactly why I need you around." Then, more serious:

"Tell me about this world." The man nodded.

"After the Coming, everything split." He held up three fingers.

"First—the Strikers. Former military, police… order through force."

"Second—the Speakers.

Those of Nadjin. They call themselves resistance."

"And then… the rest of us." He gestured to the group.

"Survivors."

"The Strikers control food. Supplies. Territory." "They starve everyone else into submission."

A man with a scruffy beard spoke up from the edge of the group.

"I used to be one of them." A pause.

"They left me behind." He looked down.

"The world changed after the Coming." He swallowed.

"We were attacked." "Something else."

"We call them the The Absent… because they don't care about life."

The shrouded man cut in sharply.

"Everything in nature would kill you if given the chance."

He stepped forward.

"But the Absent… are different." A beat.

"They came to do the same thing." "To all of us."

"They've been here for years." Cracks in the Group.

Kailee spoke, almost to herself:

"If my father were here… he'd try to control everything."

A bitter edge entered her voice.

"And he wouldn't listen to anyone. Not even me." She looked up.

"If your own family doesn't listen… why would anyone else?"

A younger girl spoke quietly:

"I became an orphan after the Coming." "They just… left me."

Styler echoed the word.

"Orphan…"

He touched his arm unconsciously.

Invisible scars.

The kind no one sees.

Another girl snapped suddenly:

"Just stop. All of you."

The group glanced at her—but didn't react.

They understood.

Everyone was breaking differently.

Then—

A pale girl with vibrant green hair and matching eyes stepped forward.

She smiled lightly.

"Maybe we all need someone else to raise us." "Maybe we're all related now."

She tilted her head playfully.

"If that's true… you all owe me years of birthdays and holidays."

A few people smiled. Just a little.

A Moment of Humanity. Kailee shifted the tone.

She looked at the girl gently.

"Your eyes… your hair—they're beautiful."

"They probably change how you see the world." Styler added, thoughtful,

"Long ago, people with blue eyes explored oceans and skies."

He nodded toward her.

"Your green eyes… they feel like growth. Like life." "You might understand things like agriculture…

ecosystems."

He glanced at her hair.

"And your hair—maybe even materials. Textures.

Fabric."

The girl blinked.

Surprised.

For a moment—

The world didn't feel broken.

Exhaustion Time slipped.

Unnoticed.

Darkness settled deeper into the shelter.

Styler exhaled heavily.

"This is too much." He rubbed his face. "I just need to sleep." A weak smile.

"Reset my head."

"Everything will make sense when I wake up." He lay back.

Closed his eyes.

The Dream

He was eating.

A small meal.

Barely enough.
Then—
A memory.
His mother's voice.
Soft. Familiar.
"Better eat before it goes cold… Or it goes old…
It looks good—Or so I'm told…"
The warmth lingered—Until it shattered.
Reality Returns.
Noise.
Panting.
Shouting.
Styler's eyes snapped open.
The shelter was alive with urgency.
A girl—pregnant—was in labor.
People gathered.
Voices overlapped.
And then—
A whisper.
"Everything starts with life…"
Moments later—
A cry.
A newborn.
Green hair.
Bright.
Natural.
Green eyes.
Alive.

12
TURMOIL

Somehow, The Rail had found its way into the bay.

Its darkened exterior blended seamlessly with the ocean floor, making it nearly indistinguishable from the surrounding abyss. Only intermittent flickers of submerged advertisements pulsed along its hull—ghostly light beneath the water. The underwater cargo trains nearby sat still, silent, as if the entire system had simply… stopped.

A World in Ruin.

At daybreak, above ground, a torn flag hung from a fractured structure.

Once, there had been many.

Now, only a few remained—relics of a world that no longer existed.

One still fluttered weakly: a holographic emblem of technology. It shimmered with fragmented images—computers, shuttles, vehicles—all arranged in a circular pattern, interconnected. A symbol of unity through innovation.

Now, it felt like a memory.

Supplies were scarce. Almost everything was.

From the underground shelters, survivors emerged cautiously, searching for food, for anything that could sustain them.

The Encounter.

Styler stood with a small group of young men in the dim light of early dusk.

A group of tanned, hardened men approached.

Their leader stepped forward, eyeing Styler's worn white shirt.

"Different colors," he said slowly, "but maybe we think the same."

He moved closer. "What's your name?" "Styler."

"Skyler?"

Styler exhaled. He was used to this.

"Styler."

The man smirked.

"Skyler." Frustration crept in. "It's Arlenish."

The man laughed harshly.

"That's the worst name I've ever heard." Another young man nearby muttered,

"Never heard it before." The leader leaned in.

"Some cultures say a name defines you. So what does yours mean?"

Then suddenly—He barked:

"JECHA!"

The other men surged forward, grabbing Styler and those around him.

Kailee's Intervention. "CHA SEN!"

Kailee's voice cut through the moment—sharp, commanding.

The men froze.

Immediately.

Styler's thoughts raced.

That means stop… Nadjin… Kailee knows Nadjin?

The leader turned toward her, anger rising.

"You have no right to speak my language," he snapped in Nadjin.

"No right to use it. To command with it. To steal it." Kailee stood firm.

"It was given to me," she said. "I respect it."

A strange wind rolled in—low, persistent, unnatural. The men grew uneasy.

"Just this time," the leader muttered.

He glanced toward the mist.

"There are worse things than me out there." They left.

Quickly.

The Strikers Arrive.

From the shadows emerged a force—organized, overwhelming.

A squadron.

Hundreds strong.

They wore black—complete, consuming black—broken only by glowing holographic insignias on their shoulders and chest.

Their uniforms were engineered: hooded jackets with integrated gloves, masks, and folding compartments that collapsed into themselves like origami.

They were known as the Strikers.

And they owned the night.

Zip restraints snapped into place.

Sonic weapons discharged in low, concussive bursts.

"They're just hungry," one soldier muttered.

Another raised his weapon.

"I live for this."

"They're defenseless," came a quieter voice.

Civilians scattered, hiding behind broken structures.

The Strikers advanced toward one of the last remaining food supply points—a grocery store barely holding together.

One of them removed his uniform layer, stepping forward to speak with civilians as if he were one of them.

A tactic.

A manipulation.

Chaos Escalates.

Styler and Kailee pushed through the confusion.

"Move! Clear out!" Strikers shouted.

Then—A shift.

A disturbance.

Something approaching.

"They're here!" a Striker screamed. "The Absent!"

The Attack.

A sound tore through the air—metal screaming, bending, dissolving.

Massive.

Unseen.

Unstoppable.

The Absent had risen—from the ocean floor.

Buildings shuddered.

Collapsed.

Styler and Kailee didn't wait to see them.

They ran.

Behind them—

People tied by the Strikers couldn't escape.

Red targeting lights scattered across the ground.

Circular projections marked casualty zones.

It wasn't a battle.

It was an eradication.

Escape.

They fled through a decaying city—structures crumbling, towering ruins reduced to skeletal remains.

Dust filled the air.

Breath burned.

Eventually—

They reached shelter.

A robotic stadium—once built for holosport—now repurposed as a refuge.

Inside, they were processed quickly.

Assigned rigid foam cots.

Separated.

Styler dragged his closer to Kailee's.

He smiled faintly.

"If I outsmarted you… you're mine." Inside the Shelter.

Nearby, a woman with bandaged wounds whispered: "I might fall asleep… and not wake up."

"Don't say that," a nurse snapped gently.

Hovering beside them was a MedBot Model 433.

A transparent rectangular machine—floating, scanning, analyzing.

It gathered data without touch:

DNA from breath, skin, retina, micro-particles.

Then—

It produced personalized medicine instantly.

Treatment without delay.

Without error.

Without hesitation.

A system built after generations of medical failure. Emotional Collapse.

Styler lay beside Kailee. His skin pale.

His voice distant.

"Nothing feels right," he murmured. "It's warm… but I feel cold."

He let out a quiet laugh as her hair brushed across his face.

Then, more serious:

"Sometimes… the only time I can bear living is when I'm asleep."

A pause.

"That's going to be more true than ever now." A scream echoed across the shelter.

"HELP ME! HELP ME!"

Kailee broke. Crying.

Styler wiped her tears with his sleeve, holding her close. It took time.

A long time.

Eventually, silence returned.

Thought and Intimacy.

Kailee turned toward him. "What are you thinking?" "It's hard to explain." Styler looked into her eyes.

Deeply.

"We all have different eyes… so we must have different ears too."

Kailee smiled faintly.

"And noses?"

She stuck out her tongue.

"And tongues?"

"Stop," he said, almost laughing. "Fine."

He continued, softer now:

"We didn't choose how we're made."

"We were given two eyes… to see everything." "Two ears… to hear it all."

"One mouth… to choose our words." He paused.

"There's design in that. Math." A small smile.

"If we were perfect… our ears would come before our mouths."

"So we'd hear ourselves before we speak." Kailee smiled.

"I like your thoughts."

"I wouldn't make it without you," he said. "And I wouldn't without you."

The Quiet After.

Light filtered through broken windows.

Dust floating in it like suspended time.

"We're going to have a long life," Kailee whispered. "Children… a future…"

A fragile smile.

"One day… I'll have to push you in a hearse." She held him tighter.

"This is our story." "Our love story."

"I'm tired," she said softly.

She drifted to sleep.

Final Reflection.

Styler stayed awake a little longer.

Thinking.

About sight.

About perception.

About how the body works like machinery—balanced, imperfect, intentional.

Then quietly:

"Close your eyes."

"When you open them… the world will give you something new."

He lay back.

And slept.

Closing Theme

We are always pulled back.

To the ground.

To rest.

To people.

To meaning.

No matter how far we go—We return.

13

DEEP

In a glimpse backward—whether memory, history, or something closer to myth—it began all at once.

Not one volcano.

Not a chain reaction.

All of them.

The Awakening

Across the planet, volcanoes erupted simultaneously violent synchronized, impossible.

Ash choked the skies.

Darkness spread like a curtain being pulled over the world.

But it wasn't just ash.

From the eruptions came something else—molten masses hurled into the atmosphere.

They didn't fall like rock.

They formed.

As they traveled through the air, the heat bled away. The molten shapes cooled, hardened, and changed.

Limbs emerged.

Structure followed.

Gravity pulled them down—not as debris, but as something alive.

They struck the earth in violent succession.

The sky gave only darkness.

The ground gave only impact.

The First Giants.

At the base of one volcano, the earth itself began to move.

A creature—mountain-sized—forced its way upward.

Not born.

Forged.

From the crater's peak, metallic forms climbed out—colossal, unnatural.

The volcano became a factory, producing them in waves.

Hundreds.

Each one hardened by the air, taking shape as it descended—solid, weaponized, alive.

The Spread.

The ground fractured under earthquakes triggered by the eruptions.

From beneath the surface—and from the skies above—they came.

Land.

Air.

Sea.

An invasion not from beyond—but from within.

Beneath the Ocean.

Far below, in the southern ocean, another eruption began.

A fissure opened along the seabed. Molten metal pushed upward through it.

Water rushed in violently, colliding with the molten flow.

Steam exploded outward.

The metal hardened—but not weakened.

It became denser.

Stronger.

Sharper.

From the fissure, an appendage emerged.

Arm-like.

Weaponized.

Then a body followed—upright, distorted, metallic.

A figure resembling life, but stripped of anything human.

The ocean tried to suppress it—cooling, compressing, resisting.

It only made them stronger.

More followed.

Then more.

Then hundreds.

Then thousands.

They broke through the ocean floor in clusters, spreading fissures as they rose.

Steam clouds concealed them, but couldn't stop them.

They came not with purpose, but with instinct.

No empathy.

No restraint.

No understanding of creation—only extraction.

They had never learned to build.

Only to take.

They were children of the planet's core.

And they communicated without sound. The Absent Rise.

In the present—

The gray metallic creatures—now known as the The Absent—covered the world.

They moved through cities like a force of nature.

Gray had once been neutral—used in vehicles and structures to withstand environmental stress.

Now it was weaponized.

The Absent attacked everything that resembled structure, function, or order.

Robotic construction units climbing skyscrapers were torn apart mid-operation.

Buildings collapsed in chains.

Entire districts vanished under their advance.

Underground Survival

Styler and Kailee remained hidden in a depot—one of many shelters scattered beneath the city.

The world above felt disconnected now.

Distant.

Unreachable.

Food dwindled.

Eventually, they survived on a single resource: Urgan flour—a wheat-like substance.

At first, it was enough.

Kailee adapted quickly—finding new ways to prepare it, to stretch it, to make it feel like something more.

But even that began to run out.

Scarcity changed them.

It sharpened thought.

Forced creativity.

Stripped away excess.

Styler looked at her.

"We have to find more food… and water."

Return to the Surface

They climbed out of the underground.

Carefully.

The world above had changed again.

Light bent strangely through the air—sharp, angular, almost alive.

Not fireflies.

Something worse.

Towering figures stood in the distance.

Molten.

Metallic.

Unreadable.

This was Styler's first true glimpse of the Absent.

The Escape

He didn't hesitate.

Sliding across the hood of a nearby vehicle—a Rastor 510—he jumped into the driver's seat.

Kailee followed.

The engine roared to life.

"Where did you get your license?" she shouted.

Styler smirked.

"Who said I had one?" "You want me to drive?" "We don't have time." They sped through the city.

Around them, metal giants tore through buildings—toppling entire structures like they were made of paper.

The Strikers' Resistance.

Ahead, a small team of Strikers moved toward a damaged skyscraper.

They worked quickly—placing thermite charges at the base.

"CLEAR!"

They scattered. Moments later—The explosion hit.

The building collapsed inward, crashing down onto an advancing creature.

It slowed it.

Not stopped it.

Dust and debris filled the air.

The world shook again.

Collapse.

Styler pushed the vehicle faster.

But the roads were failing.

Rubble piled up faster than they could navigate.

They passed a pair of Strikers catching their breath. One joked:

"They've destroyed everything… we'll have to rename all the satellite coordinates."

Another laughed.

"Go ahead. Name something after me." Even now—humor survived.

Then—

The car hit debris.

Hard.

It stopped.

Silence, for a moment.

Styler looked ahead.

Then at Kailee.

"We're on foot now."

And the world waited for them—Broken, burning, and still moving.

14

UNDERGROUND

The underground depot had grown quieter with each passing day.

Not peaceful. Restless.

The dim cavern walls glowed faintly from scattered Carbocandle light, throwing long shadows over exhausted faces. People paced.

Whispered.

Argued softly.

Survival without direction was beginning to rot morale from the inside out.

Styler stood watching everyone.

Then finally said:

"Just waiting around… doing nothing… is a mistake." Several people looked toward him.

The shrouded man leaned against a concrete pillar, arms folded.

"So what's your plan?" he asked.

Styler thought for a moment. "We need structure."

He looked around the depot.

"Food. Water. Repairs. Defense. Information. We keep acting like people hiding underground instead of people trying to survive."

An older woman muttered, "Easy to say."

"No," Styler replied calmly. "Easy is sitting here starving until something finds us."

The room fell quieter.

Even the younger children stopped talking.

Building Purpose.

Styler walked toward the salvaged robot parts the child called Trucker had been rebuilding.

"We already have skills here," Styler continued.

He pointed around the room.

"Mechanics.

People who know Nadjin.

Medical experience.

Farmers.

People who know how to ration.

People who know how to organize."

The shrouded man smirked.

"You sound like a leader now." Styler shook his head immediately.

"No. Leaders get people killed if they think they matter more than everyone else."

A pause.

"We need people useful at different things." Kailee smiled faintly from nearby.

That sounded like him.

Organizing the Underground.

Over the next several hours, Styler pushed people into motion.

Not through authority.

Through purpose.

Water collection teams were assigned.

Others organized the remaining food stock.

A group began cataloging supplies recovered from outside.

Even Shannon got involved, cheerfully helping despite the tension.

"Look at us," she laughed while stacking containers. "We're basically civilization again."

"Barely," someone answered.

Still—

It helped.

Movement gave people hope.

Trucker's Discovery.

Near the edge of the depot, Trucker continued rebuilding the fallen cleaning robot.

Wires sprawled around him.

The boy hardly blinked while working.

Styler crouched beside him.

"How's it going?"

Trucker tightened a connection.

"Almost."

Sparks flickered.

One of the robot's optical sensors suddenly glowed faint blue.

People nearby turned immediately.

The child grinned proudly.

"Told you."

The damaged robot twitched.

Not fully functional.

But alive enough to move.

The Philosophy of Survival.

The older woman who had spoken to Styler earlier approached again quietly.

"You know what your problem is?" she asked.

Styler looked at her cautiously.

"You still think the future is waiting for you." That sat with him.

Heavy.

She gestured toward the people around them.

"These people already lost theirs."

Styler glanced around the underground.

People sleeping beside walls.

Children clinging to strangers.

Others staring blankly into candlelight.

Maybe she was right.

Kailee and Styler

Later, away from the noise, Kailee sat beside Styler near the underground waterfall system.

The water echoed softly through the cavern.

For a moment, it almost sounded peaceful.

Kailee leaned against him.

"You're changing."

Styler stared into the water.

"I think this place changes everyone." "No," she said quietly.

"You're becoming who you were supposed to be." That unsettled him more than comforted him.

The Signal Suddenly—

A noise echoed through the underground.

Not mechanical.

Not human.

The partially rebuilt robot jerked upright.

Its damaged speaker crackled violently.

Everyone froze.

Then—Static.

Fragments of sound. A transmission.

"…remaining survivors…" "…northern sector…" "…transport…"

"…safe district…"

People rushed toward the robot. The signal broke apart again.

But one phrase came through clearly:

"—there are still cities standing."

Silence hit the underground harder than panic ever had.

Hope.

Real hope.

The shrouded man stepped forward immediately.

"That transmission shouldn't exist." Trucker looked up proudly.

"It exists now." Styler's mind raced.

If there were surviving cities… Then there were still factories. Still systems.

Still people fighting back. Or worse—

People controlling what remained.

A Decision

Kailee looked at Styler.

"You're thinking about leaving already." He didn't answer immediately.

Because she was right. Finally—

"We can't stay buried forever." The shrouded man overheard.

"If you go looking for surviving cities," he warned,

"you're going to find war before you find hope." Styler looked toward the cavern ceiling. Toward the world above them.

"Then we find out what's left anyway." Kailee tried to lighten the mood.

"We all make mistakes," she joked weakly. "And we're definitely going to make more."

Styler didn't laugh.

"This is a waste of time."

The words came out flat, exhausted.

Outside, the sky split open.

Heavy rain hammered the ruined city above them, every droplet echoing through broken drainage tunnels and fractured concrete. Water began leaking steadily into the underground shelter.

People scrambled to move blankets and supplies.

The depot's poor drainage quickly turned the floor slick and muddy.

Kailee stared upward as thunder shook the ceiling.

"Did I make that happen?" she asked quietly.

"The weather changes and somehow I feel like it was me."

Then she caught herself and shook her head.

Funny how people blamed themselves for everything.

As if every storm needed a witness.

Suddenly—

Shouting erupted near the entrance tunnels.

"STRIKERS ARE COMING!"

Panic exploded through the underground. People grabbed children.

Others rushed for supplies.

A few simply froze.

Styler immediately stood.

"We gotta move!" More voices echoed:

"Strikers incoming!"

The City Above.

For a split second, the city lost power. Darkness swallowed the skyline.

Then—

Light returned.

Emergency systems surged alive across the metropolis.

The city's hidden backup battery systems had activated, powering entire districts at once.

But the Absent had noticed.

The metallic creatures swarmed toward the energy infrastructure like starving predators. They fed on dense materials, on metals, on power systems themselves. The more advanced the structure, the more violently they attacked it.

Buildings shook.

Far above, AI defense turrets opened fire.

The automated turrets generated ammunition directly from compressed atmosphere, pulling air inward and condensing it into dense projectiles.

Near-infinite ammunition.

Near-infinite desperation.

Through the underground hallways came distant whispers in Nadjin, bouncing eerily through the tunnels.

Then—Gunfire.

Not against the underground.

Against the Absent.

The Strikers were fighting them.

Survival Instinct.

Kailee moved toward the younger children.

"We need them asleep," she said softly.

"If we stay awake all night, we won't survive."

Even animals knew not to exhaust themselves in darkness.

Human civilization had once depended entirely on daylight.

Until machines replaced the night.

Styler leaned against the wall.

"I know we can't change the sun and moon," he murmured, "but maybe we're not getting through this."

Kailee looked toward the dwindling food supplies.

"We're going to need more food." Styler made his decision immediately. "I'll go."

The Supermarket.

The supermarket was still guarded. Still fortified.

Still treated like sacred territory.

Styler slipped inside during the confusion.

Outside, an elderly woman approached the entrance weakly.

"I haven't eaten in a week," she pleaded.

A Striker shoved her backward with his atmospheric rifle.

"You better start farming."

Another guard stepped closer, almost amused.

"You want food?" he mocked. "Beg for it."

He raised the rifle near her face.

"I want you to cry for it."

Styler watched from inside, horrified. Then chaos erupted.

A group of armed Speakers stormed the supermarket perimeter.

Gunfire exploded instantly.

One Striker shouted into a communication device: "Tact-com! Build me four walls!"

Portable barricades deployed around the storefront. Inside, atmospheric rifle rounds tore through civilians.

Two people dropped instantly.

Styler reacted without thinking.

He grabbed packaging from raw meat containers and wrapped one wounded person's bleeding leg tightly with the clean plastic side inward. Using the cold packaged meat externally to weigh the wound down, he improvised pressure and cooling simultaneously.

Then—

Blood splashed nearby.

Styler smeared some onto himself.

Dropped to the floor.

And stayed silent.

Bodies screamed around him.

Gunfire.

Shattering glass.

Smoke.

Then quiet enough.

He ran.

Cornered.

Outside, Styler reunited with Kailee and one of the younger underground members who had come searching for him.

They fled through collapsing streets until—Dead end.

An alley.

No exits.

Several Strikers entered from the opposite side.

Weapons raised.

Kailee whispered to the younger member: "Don't trust them. Not one of them."

One muscular Striker leveled his rifle toward the younger boy first.

Then toward Styler and Kailee. "Leave us alone," Kailee snapped. "My family helped build this city."

Before the confrontation escalated—gunfire erupted overhead.

The Speakers had taken rooftop positions.

They opened fire directly onto the Strikers while intentionally avoiding unarmed civilians below.

The Strikers immediately lost formation.

"MOVE!" Styler shouted.

They escaped toward a nearby Rastor 510.

Styler slid across the hood, jumped into the driver seat, and activated the vehicle.

Kailee and the younger underground member piled inside.

The Collapse.

As they drove back through the city, civilization visibly unraveled.

People looted food depots openly.

Water supply stations were being torn apart.

Supply points burned.

The Absent continued advancing through districts, toppling buildings and spreading panic faster than the storms themselves.

By the time they returned underground, they had barely gathered anything useful.

Then—

The ceiling shook violently.

Dust rained downward.

Half the depot ceiling collapsed inward.

The lights flickered.

Darkness swallowed the room for several seconds.

People screamed.

Children cried.

Kailee shouted over the panic:

"We can't stay here!"

Styler stared into the dust clouds forming through the candlelight.

"What do you want to be known for in history?" No one answered.

The Plan

The shrouded man stepped forward.

"There's a place I know." Everyone listened.

"A deserted military base.

Supplies.

Equipment.

Technology."

His expression darkened.

"It gives us a chance." "But?" Styler asked.

"It's dangerous. And heavily patrolled." The shrouded man looked directly at Styler. "I can get you there."

Kailee immediately understood.

"You want us to distract the Strikers." Styler nodded.

"You and the others pull attention away. Me and him get inside the base."

Silence followed.

Then Kailee quietly turned toward the shrouded man.

"Take care of him." A pause.

"Take care of Styler."

Then they disappeared into the ruined city.

Melissa's Betrayal.

As Kailee organized the remaining underground members, Melissa slipped away unnoticed.

Unlike everyone else, she walked openly through the streets.

Fearlessly.

She approached three Strikers standing watch.

One had red hair protruding from beneath his hood. Another had striking green hair.

Melissa smiled brightly.

"I've got something to tell you." A beat.

"And you're probably not going to like it." She whispered to them.

Slowly, their expressions hardened.

Something about the conversation carried immediate weight.

The Speakers.

Following Nadjin graffiti markers through ruined districts, Kailee eventually reached a shattered skyscraper filled with hundreds of Speakers.

Some armed.

Some silent.

All watching.

A Nadjin storyteller sat among younger members, speaking softly:

"People walk into houses through doors like mouths…"

Kailee listened carefully before stepping forward.

"Den je Nadjin," she said.

Mind if we talk?

The room shifted immediately.

A man laughed darkly.

"So you learned our language." Another stepped forward aggressively.

"You think learning a language makes you one of us?" The accusations came rapidly.

"You steal from native speakers." "You stain the language."

"You don't belong here."

One Speaker grabbed Kailee by the shirt.

"Come with me."

Another interrupted sharply:

"Let her speak." Kailee steadied herself. "I am on your side."

Then conversations shifted rapidly around the room.

The Absent were overwhelming entire districts.

The unthinkable possibility emerged:

The Speakers and Strikers might need to unite.

One man shouted:

"This is suicide!"

Others murmured agreement.

Fear spread quickly.

Then—

Strikers stormed the building.

More than twenty entered the skyscraper seemingly out of nowhere.

Kailee's stomach dropped instantly.

Melissa.

She realized what had happened.

The Speakers reached for weapons.

The Strikers blocked the boulevard outside.

One guard stepped forward with a grin. "You're not getting through."

He raised his gun slightly.

"I've got a weapon." A pause.

"So are you willing to die trying?"

15
HAUL

Beyond the shattered outskirts of the city, stretching across the horizon, lay a massive industrial field.

Factories.

Hundreds of them.

Some dormant.

Some still operational.

Most coughing smoke into an already wounded sky.

At the center stood a sprawling solar-driven recycling and atmospheric reduction facility.

Even from a distance, Styler could hear it breathing.

The pollution reduction complex pulled immense volumes of air inward through colossal intake towers.

Standing nearby felt like fighting the atmosphere itself just to inhale.

The systems extracted carbon dioxide from the environment, breaking it down into oxygen and pure carbon compounds. The excess oxygen was compressed into storage tanks for spaceflight use, while the carbon was refined into industrial-grade carbon fiber.

Airplanes made from air.

Cars made from pollution.

The irony lingered heavily in Styler's mind.

Even while humanity tried to repair the planet, Styler couldn't stop thinking about how avoidable it all had been.

Fossil fuel emissions never needed to happen.

Cars could always have been electric.

History had simply chosen convenience over foresight.

Inside the moving Rastor 510, Styler stared through the window as flocks of birds spiraled chaotically through the cloudy sky.

They weren't migrating.

They were panicking.

Their sense of direction had been disrupted by the Absent.

Animals felt it before humans did.

The Rastor entered an aging tunnel once used during the fossil fuel era as a carbon-scrubbing experimental corridor.

Dim lights flickered overhead one by one as they drove deeper. Then music crackled softly through the radio.

"There's a story in your eyes… A story in disguise…"

Styler immediately thought of Kailee.

Wondering if she was alive.

Wondering if the plan had worked.

Emerging from the tunnel, they passed untouched neighborhoods constructed in haste decades earlier.

Entire apartment complexes had been 3D printed directly onto massive concrete foundations.

Synthetic layered materials mimicked the warmth of wood while retaining industrial durability.

Farther ahead, something stranger appeared.

Buildings phased into existence.

Not constructed.

Appeared.

4D-printed architecture.

Structures manufactured across multiple points in time and inserted into the present to combat suburban sprawl.

Styler watched in awe.

Humanity had once bent time itself simply to solve housing shortages.

The Military Facility.

Eventually they reached the outskirts of the military compound.

The Shrouded Man smiled grimly.

"This is it."

They abandoned the Rastor, scaled the perimeter fence, and slipped inside through a maintenance breach.

The interior was overwhelming.

Rows upon rows of autonomous manufacturing systems operated in silence.

Computers manufactured smaller computers.

Those smaller computers produced microscopic processors.

Then even smaller ones.

Nested endlessly like technological matryoshka dolls.

Some components were invisible to the naked eye entirely, only observable through magnification screens.

Styler stared in disbelief.

This was what the Information era should have become.

Instead of endless vanity and repetitive entertainment, humanity could have devoted itself fully to creation.

In another chamber, Styler discovered enormous geological archives.

Core samples.

Mineral collections.

Botanical specimens.

Entire walls cataloging biology and geology together.

He suddenly understood Harold's obsession with collections.

Knowledge organized physically became survival.

If they ever returned home, he wanted to build something like this himself.

They moved deeper into the facility.

Ahead stretched a massive command center built for hundreds of engineers.

Now completely abandoned.

Towering holographic displays projected deep-space relay depots across known space.

Unmanned stations positioned like stepping stones between planets.

Each depot refueled spacecraft automatically.

Each new mission extended humanity farther outward. Civilization had once planned for infinity.

Farther inside rested unfinished rockets and satellites unlike anything Styler had ever seen.

"What are they building?" he whispered.

The Shrouded Man studied them carefully. "Sataforming technology."

Styler frowned.

"Large-scale satellites capable of altering molecular structures remotely," the Shrouded Man explained.

"They can evolve matter itself.

Create compounds.

Change environments across entire planets."

He pointed toward another satellite frame.

"That one appears designed to simulate stellar radiation. Artificial sunlight."

Styler stared.

"So theoretically… one could turn a mountain into

gold?"

A pause.

"Or create oceans in empty space?"

"Possibly," the Shrouded Man admitted.

"But they're incomplete." Styler sighed.

"Would've made things easier." The Shrouded Man smirked.

"Too bad."

The Timebomb.

Styler discovered a massive laminated emergency manual resting near a workstation.

Thousands of pages.

Step-by-step procedures.

Diagrams.

Failsafes.

He skimmed rapidly, mentally memorizing layouts and systems.

Nearby stood spacecraft with regenerative hull systems capable of self-repair like living tissue.

Others featured detachable multi-directional passenger modules designed for simultaneous interplanetary deployment.

Humanity had engineered wonders.

And abandoned them.

Then Styler noticed something unusual resting on a nearby table.

A dark device with transparent panels revealing swirling liquid and granular material inside.

"What's that?"

The Shrouded Man looked over casually.

"Don't touch it." A beat.

"That's a timebomb." Styler's eyes lit subtly. Useful.

While the Shrouded Man wasn't looking, Styler quietly slipped one into his side pocket.

Preparing for War.

Together they loaded ether thermite explosives, weapons, and crates of supplies into the Rastor.

Then they crossed into the aircraft hangars.

Rows of massive hybrid aircraft filled the chamber. Spacecraft. Atmospheric craft.

Ocean-capable vehicles.

All dormant.

Inside one transport vessel, Styler discovered an unusual cryogenic suspension system.

Instead of individual hibernation pods, the entire cabin could be frozen simultaneously through atmospheric regulation.

Mass suspended animation.

Efficient.

Terrifying.

The Shrouded Man hacked into one aircraft's reactor systems.

"There's more you can learn from me than you think." Styler climbed into the engineering compartment of another aircraft and immediately began rerouting systems manually.

Wires detached.

Panels opened.

Console pathways reconfigured.

The Shrouded Man watched silently for a while.

Then finally said:

"You're proof greatness exists in everyone."

Styler dropped from the emergency ladder triumphantly.

"Done!"

Suddenly—

Voices echoed through the hangar.

Laughter.

Nadjin. Dangerously close.

A group of Speakers dragged a man toward a fusion propulsion engine.

One of them shouted mockingly in Nadjin: "We got some turbulence!"

Then they threw him into the engine chamber.

Styler froze.

Another man laughed bitterly.

"Why trade for anything when you can kill for it?" The Shrouded Man whispered:

"Do I look like them?" Styler immediately bolted.

"Great!" the Shrouded Man yelled sarcastically while following.

"We can die together!"

"No one has to die!" Styler shouted back.

They burst into an unlit depot.

"LIGHTS!"

The room illuminated instantly.

And Styler stopped dead.

Rows upon rows of robotic soldiers stood dormant inside.

At least forty.

Motionless.

Waiting.

"There has to be forty of these..." The Shrouded Man grinned.

"What's the point of counting what you can't use?"

Styler desperately searched for a control system but couldn't find one.

Nearby, the Shrouded Man casually activated a nearby 4D printer.

Then—

Shouting erupted behind them.

They'd been found.

One Speaker advanced on Styler aggressively, shouting in Nadjin.

Styler blinked.

Then shrugged.

"Well... sorry for the inconvenience." Suddenly—

An atmospheric rifle materialized directly into Styler's

hands.

The 4D printer had fabricated it instantly.

Styler fired immediately.

The pursuing Speakers scattered.

"Didn't see that coming," Styler grinned.

The Shrouded Man walked past him laughing.

"Thought you'd need that in about five minutes."

Reunion.

They loaded the Rastor with everything possible.

Weapons.

Explosives.

Supplies.

Then returned toward the collapsing city.

Ahead, Strikers exchanged fire with the Absent.

The creatures communicated silently with one another.

Not separate.

One collective consciousness.

Far ahead, Kailee moved through rubble alongside several underground survivors.

Suddenly—

A hand clamped over her mouth.

She was dragged backward violently into an alley.

Kailee struggled wildly until the voice whispered: "Kailee… it's me."

She turned instantly.

"STYLER!"

Relief flooded her face.

"The plan's in motion," Styler said quickly.

Around them, the city burned.

The Absent attacked indiscriminately.

And for the first time—

The Speakers of Nadjin and the Strikers shook hands.

16
AWAKENING

Carefully and patiently, with explosives loaded into the back of the Rastor 510, Styler and the Shrouded Man drove back toward the underground shelter.

The city had transformed into preparation. Preparation for war.

Reprogrammed flying 3D printers buzzed overhead, constructing barricades and traps under the guidance of Trucker and several Strikers.

One Striker shouted as the Rastor rolled past: "Watch the traps!"

Metal frameworks unfolded from the air itself.

Explosive nets.

Collapsing barriers.

Sonic mines.

Humanity had become inventive again because survival demanded it.

Thousands of Speakers and Strikers gathered across the shattered streets preparing for battle against the Absent.

For the first time, they stood together.

The Underground joined them too.

Fear no longer separated anyone.

Kailee crouched beside crates of explosives, fastening Carbocandles to makeshift charges that could be hurled at the advancing creatures.

Nearby, the Shrouded Man distributed explosives across the avenue.

Styler looked over the battlefield.

"Is everything in place?"

"As much as it can be," Kailee answered confidently.

But Styler's thoughts spiraled.

What if I'm wrong?

What if all of this is a mistake? Kailee looked directly into his eyes. "Styler… slow down your mind." Then—

A nearby skyscraper collapsed beneath the assault of an Absent.

The impact shook the entire avenue.

Styler hit the ground hard.

For a moment, he couldn't move.

Couldn't think.

Couldn't breathe.

The pavement beneath him felt freezing.

Cruel.

Dust coated his lungs.

His heartbeat pounded violently in his ears.

Kailee knelt beside him, gripping his face gently. "You are my everything…" she whispered.

"Come back."

Trucker suddenly appeared beside them.

Without hesitation he placed the Icon A4 onto Styler's ear and activated medical mode.

A soft therapeutic frequency played.

The device monitored his vitals, calming his nervous system almost instantly.

Styler slowly steadied himself.

Then smiled weakly.

"Thanks."

The World Breaking Apart.

The Absent existed across all points in time simultaneously.

Buildings launched themselves into orbit to escape them.

Entire skyscrapers ignited hidden thrusters beneath their foundations and ascended into the sky carrying desperate survivors toward relay depots and distant colonies.

Others anchored deeper into the earth attempting to survive where they stood.

From the southern oceans, vast fractures opened beneath the water.

Sea levels shifted.

The magnetic pull of the planet itself seemed disturbed.

More Absent emerged.

Airplanes refused to land.

Some circled endlessly overhead powered by atomic engines capable of remaining airborne for years if necessary.

They waited for safety that never came.

The city became a nightmare in daylight.

Towering metallic masses slammed through skyscrapers miles high.

Glass rained downward.

Window-cleaning drones plummeted from collapsing towers onto civilians below.

A man was severed apart nearby.

Kailee froze in horror.

"Call an ambulance!"

But there were almost none left.

The Shrouded Man fired his atmospheric rifle into one of the towering Absent.

The rounds only chipped fragments away. Styler looked toward him carefully. "Ready to do something a little crazy?"

The Shrouded Man narrowed his eyes.

Then the autonomous Rastor accelerated directly toward an advancing Absent.

"Little crazy?! Did you SEE that?!" Styler grinned grimly.

"Then you're in."

Humanity Fights Back.

Styler hurled an explosive.

It detonated against an Absent's torso, tearing away an enormous chunk.

The creature collapsed. For the first time—The city cheered.

Speakers and Strikers alike erupted in celebration.

Now they knew the Absent could be damaged.

Could be killed.

The avenue erupted into warfare.

Atmospheric rifles cracked through the air.

Makeshift catapults launched explosives.

AI defense systems tracked targets automatically.

One Absent fell into a trap ditch.

Another crushed a Striker beneath its weight.

Still they kept coming.

Enough to fill the horizon.

Styler murmured:

"Everything takes time." Kailee suddenly shouted:

"IT'S TIME!"

The Underground launched explosive volleys while Speakers and Strikers advanced together through smoke and debris.

Six. Seven.

More Absent fell.

One Striker screamed over the chaos: "This is what we die for!"

"MOVE FORWARD!"

United forces charged.

The Explosion.

Amid the battle Styler heard a voice guiding him.

"Keep going. You'll find her." Through the smoke he saw Kailee. "Kailee… I found you…" Then—

A massive explosion detonated nearby.

Styler collapsed.

His lungs tightened.

Sweat drenched his shirt.

Everything faded.

"STYLER!"

Kailee's voice pierced through the ringing.

Slowly his eyes reopened.

He saw her.

And somehow fell in love with her all over again.

"I'm fine," he whispered weakly.

In the chaos, something slipped from his pocket unnoticed.

The timebomb.

Nearby, a Speaker activated "The Screen," projecting it horizontally across the avenue like a barrier wall.

For several seconds it held.

Then the Absent tore directly through it. Sirens blared across the city.

"AUTOMATED DEFENSE SYSTEM." "CLEAR AND EVACUATE."

A distorted holographic commercial flickered across the smoke-filled sky.

The advertisement malfunctioned into a nightmare image of AI defense systems executing people automatically for "unauthorized entry."

Even advertisements had become weapons.

At a distant skyport, civilians desperately fled toward aircraft.

Suitcases followed automatically behind them like obedient pets.

Then an Absent attacked one of the support towers.

The airport partially collapsed.

Planes and passengers fell screaming from the sky.

The Shrouded Man stared upward in fear.

"The Absent encompass all points in time…" Kailee's Collapse

Styler suddenly checked his pockets.

"Who has the timebomb?!"

Nearby, Kailee crouched beside several Speakers.

Then something changed.

Her body trembled violently.

"Styler…" she whispered shakily.

"Have you ever felt something that couldn't possibly be natural?"

She collapsed to her knees laughing uncontrollably.

Not her own laughter.

Her movements seemed manipulated.

Like someone else was inside her.

Styler froze in horror.

The Shrouded Man whispered rapid Nadjin phrases into her ear desperately.

Gradually—

Kailee regained control.

She opened her eyes slowly and reached for Styler. Overhead, a rocket descended through the clouds.

Its landing thrusters adjusted automatically before impact.

Then compartments burst open.

Weapons.

Supplies.

Exo-suits activated through chemical sweat reactions.

AI turrets.

Medical kits.

Robotic combat units.

The city had become a full-scale warzone.

The shadows themselves seemed alive.

The Absent emerged from darkness between buildings.

Every street became a battlefield.

Collapse.

Bridges beneath the ocean floor collapsed.

Medical trains derailed.

Unnatural weather spread through the city.

"It's caustic out," automated broadcasts warned.

Styler and Kailee fled through ruined streets until they found refuge inside the wreckage of an old time travel venue.

But even there—

Another Absent approached.

Roadways shattered beneath its weight.

Underground landfill transit systems collapsed inward.

Entire infrastructure networks vanished beneath the city.

Kailee threw an explosive. They ran.

An atomic passenger aircraft crashed into the ocean nearby.

Speakers marked themselves with ash across their uniforms for identification.

One of them said coldly:

"Some people are worth more dead." Kailee suddenly shoved Styler aside.

An Absent crashed where he'd stood moments earlier. His hand slammed against debris, slicing deeply open. Kailee examined the wound quickly.

"I should've been a doctor," she joked softly.

Styler stared blankly at the blood covering his hand. Finally—

Speakers.

Strikers.

The Underground.

Everyone gathered every remaining explosive for one final assault.

Launchers fired.

Impact grenades detonated.

Carbocandle explosives erupted across the avenue.

One by one the remaining Absent collapsed.

Then silence.

A roar spread across the survivors.

Not celebration.

Relief.

Sanctity.

Everyone had survived.

The battle was over.

But they still remained trapped in the future.

The Shrouded Man approached Kailee quietly. "They were wrong about you."

"They said you wouldn't survive."

A nearby Striker approached Styler holding something carefully.

"I believe this is yours." The timebomb.

Styler exhaled sharply with relief.

"Thank you."

Days later, while others recovered, Styler began reprogramming surviving construction robots and Farsers exactly the way Harold had taught him.

He linked directly into their systems.

Rebuilt them.

Repurposed them.

Then finally smiled. "I've got a surprise."

Together with the machines, Styler began constructing an entirely new neighborhood from the ruins.

Not hiding anymore. Building.

Kailee watched the robots tending crops nearby and laughed softly.

"It's funny."

"Robots grow the food…" She smiled at Styler. "But robots don't eat."

17
DARKNESS

The gravel crunched beneath the students' shoes as the class moved together along the familiar outdoor trail.

The morning air carried the sharp scent of cold stone and damp earth after the recent snowfall.

Above them, automated maintenance drones drifted silently between the trees, trimming dead branches before they could fall across pathways.

Styler walked quietly near the back of the group.

Everything around him looked the same.

But nothing felt the same anymore.

His teacher repeated his favorite phrase with calm conviction:

"It's not the roads you walk on. It's the ones you make."

The class continued forward, but the words struck Styler differently now.

He had walked through ruined futures. Watched cities collapse. Seen humanity nearly disappear. And somehow—

He had returned to ordinary life.

Sean glanced back at Styler while walking.

"You look terrible." Styler smirked faintly. "Thanks."

"No seriously," Sean continued. "You look like you fought a war."

Styler froze for half a second.

Then laughed under his breath. "If only you knew."

The students eventually reached the grassy clearing where classes were often held outdoors.

Some sat cross-legged immediately.

Others sprawled lazily across the grass.

Styler remained standing for a moment, staring upward through the trees as wind shifted the branches overhead.

He kept expecting to hear sirens.

Or collapsing buildings.

Or the metallic screeching of the Absent.

But there was only birdsong.

The Teacher noticed his distraction. "Something on your mind today, Styler?" Styler hesitated.

How could he possibly explain any of it?

That he had lived through months no one else remembered.

That he had watched the end of the world and somehow returned in time for school.

"That obvious?" Styler asked quietly.

The Teacher smiled.

"The mind usually leaves footprints somewhere." A few students chuckled.

Styler finally sat down.

Jessica was absentmindedly drawing lines through the dirt beside her shoe.

The Teacher noticed.

"Jessica, what are you making?" She shrugged.

"Nothing really."

The Teacher nodded thoughtfully.

"Interesting thing about people. We say we're doing nothing, but the mind almost never stops creating."

He looked around the circle.

"Even destruction is creation in reverse." Styler looked down at his hands.

The same hands that had carried explosives.

Rebuilt machines.

Held Kailee while cities burned.

Now they rested quietly in his lap like none of it had happened.

Dave leaned toward Styler from nearby.

"You disappear for like two days and come back looking like a cave survivor."

Styler laughed softly.

"Felt longer than two days." Dave smirked.

"You miss another assignment?" Styler looked at him.

"If that's the worst thing that happened, I'm doing pretty

good."

The Teacher stood and slowly paced around the circle.

"You know," he said, "most people spend their whole lives trying to become someone else."

He gestured toward the students.

"But eventually life corners you."

"And when it does, all you really find out is who you already were."

Styler listened carefully.

For once, he wasn't overthinking the words.

He simply let them sit with him.

A breeze swept through the clearing.

Leaves spiraled down around the students.

The world felt peaceful again.

Almost unnaturally peaceful.

Then Kailee appeared walking up the gravel path toward the class.

Late.

Completely unapologetic.

The moment Styler saw her, something inside him settled.

She smiled immediately after spotting him.

"There you are." Styler smiled back. "There you are."

The Teacher folded his arms knowingly.

"Well," he said dryly, "I can already tell nobody's paying attention to me anymore." A few students laughed.

Kailee slipped into the circle beside Styler and quietly leaned close enough for only him to hear.

"Still feels weird being back." Styler nodded slightly.

"Yeah."

After a pause he added:

"Sometimes I still think we're there." Kailee looked toward the sky.

"So do I."

For a moment neither of them spoke. Then Kailee whispered:

"You know what scares me most?" Styler looked at her.

"What?"

"That eventually," she said softly, "this is going to start feeling normal again."

The thought hit him harder than expected.

Because part of him already felt it happening.

The future they survived was becoming memory.

Like a dream fading after waking.

The Teacher suddenly addressed the class again. "One final thought before we conclude." Everyone looked up.

"History remembers invention.

It remembers wars.

It remembers leaders."

He paused.

"But what actually changes the world…" His eyes shifted briefly toward Styler.

"…is often just people refusing to give up on each other."

The class fell unusually quiet. Styler lowered his eyes.

Kailee gently slipped her hand into his beneath the grass where no one else could see.

Class concluded shortly after.

Students scattered down different paths through the city outskirts.

Styler and Kailee remained behind walking slowly together.

Neither seemed in a hurry anymore.

Far above them, The Rail glided silently across the skyline like a dark ribbon of moving light.

Construction drones climbed skyscrapers in the distance.

The city lived.

People laughed.

Traffic moved. Life continued. Styler finally spoke.

"Do you ever think about them?" Kailee already knew who he meant. Shannon.

Trucker.

The Shrouded Man.

All the people stranded in the future they had left behind.

"Every day," she admitted.

Styler nodded.

"Me too."

Kailee stopped walking and turned toward him fully.

"But maybe that's the point." "What is?"

"That they mattered." She smiled sadly.

"If we remember them… then they existed." Styler looked out across the city.

A civilization still standing.

A future not yet broken.

Then quietly he said:

"Maybe standing the test of time isn't about surviving forever."

Kailee tilted her head. "Then what is it?" Styler looked at her.

"It's leaving something behind worth remembering."

18
ONE MORE YEAR

It was one of those brutally bright summer days where the sunlight didn't simply warm your skin — it punished it.

The kind of day that promised a deep red burn by evening.

Yet somehow, the heat felt meaningful.

Important.

Today mattered.

Styler's graduation ceremony had finally arrived.

So had everyone else's.

The school, as always, remained modest about it all.

Decorations were minimal — a few holographic ribbons, several faded banners, and a projected congratulations sign that occasionally flickered.

Administrators didn't consider this a major graduation year.

But to the students sitting there? It felt enormous.

A finish line.

A beginning.

Rows of students sat beneath the blazing sunlight dressed in formal uniforms that trapped heat mercilessly.

Styler sat among them quietly.

A long exhale escaped him.

The weight of the year seemed to leave his body all at once.

Harold noticed immediately from the audience.

So did Kailee sitting beside him.

Both cheered loudly whenever Styler's row was mentioned.

Three students were called first. Then—

"Styler."

His heart skipped.

He stood and walked toward the stage.

The applause sounded strangely distant to him, almost dreamlike.

At the podium, a sleek certification device projected a hovering interface in front of him.

"State confirmation," it instructed.

Styler leaned toward the microphone. "Conferred."

The machine emitted a soft tone and instantly generated his graduation certification.

The audience applauded.

On impulse, Styler removed his cap and tossed it toward Harold.

Harold barely caught it.

The crowd laughed and cheered even louder.

Then the ceremony paused for the keynote projection.

A holographic figure materialized above the stage.

It was inventor John Williamson — or rather, a preserved recording of him.

The hologram smiled warmly.

"It is a great day for everyone," the recording began. "But especially for this year's graduates."

"You have all been given an opportunity." "A difficult one."

"Perhaps the most difficult opportunity any generation receives."

"But difficult paths often lead to meaningful places." He paused briefly.

"May the stars and heavens reveal your paths in the years ahead."

Styler leaned slightly toward the student beside him.

"What do you think he means?" The student shrugged.

"No idea."

As the ceremony concluded, students scattered toward families and caretakers waiting beyond the seating area.

One classmate stopped Styler.

"So what are you going to be when you grow up?" Styler thought for only a moment.

"I'm going to be what's needed." Harold's Warning.

Later that evening, Styler, Harold, and Kailee returned to Harold's house in the Morrison Mod 5.

There was already food spread across the kitchen table waiting for whoever stopped by.

Simple food.

Comfort food.

The kind Harold liked.

But during dinner, Harold quietly stood and disappeared into the other room.

Styler noticed something wrong immediately.

He followed him.

Harold stood near the wall gripping the edge of a counter.

Pain crossed his face.

Real pain.

Not the kind he usually joked through.

Styler's stomach tightened.

"How bad is it?" Harold smirked weakly. "It's over kind of bad." Then after a pause:

"But also just starting." Styler looked shaken.

Harold motioned for him to sit.

"So, my future engineer…" His voice softened.

"When you feel alone… or overwhelmed… or like life is unfair…"

He gestured vaguely around the room.

"Look around."

"At the home you didn't build yourself." "The food you didn't grow."

"The inventions you didn't invent." "The roads you didn't pave."

"The people who labored before you so you could simply exist comfortably."

He smiled faintly.

"Don't think about the money attached to things.

Think about the effort."

"The sweat."

"The sacrifice." "The time."

Harold continued quietly:

"That's what kept me at peace my whole life." "Realizing how much of life was built by people trying to help someone they'd never even meet." Styler listened silently.

Without realizing it, he was memorizing every word. The Morning Everything Changed.

That night, both Styler and Harold went to sleep early.

The house became quiet.

Dark.

Still.

Then morning came.

And something felt wrong immediately.

Too quiet.

Harold was always awake before him.

Always.

Styler sat up slowly.

"Harold?"

No answer.

He walked into the hallway.

"Harold?"

Still nothing.

A cold feeling spread through him.

He moved faster toward Harold's room. "Har… Harold?"

He stepped inside.

Then froze.

Harold lay motionless in bed.

Pale.

Stiff.

The life had completely left him.

Styler's breathing broke apart. "Harold…"

No response.

"Harold!"

His voice cracked into panic.

Outside, an AutoCop levitated toward the house after detecting abnormal biometric readings and elevated distress signals.

Styler stumbled toward the doorway as it arrived.

"I went in and he wasn't moving…" The AutoCop processed silently.

Then replied:

"Deceased."

"Likely cause: age-related deterioration." The words destroyed him.

Not because they were cruel.

Because they were mechanical. Simple.

Like Harold's entire life had been reduced to a diagnosis. Styler left the house almost immediately.

He didn't know where he was going.

Only that he couldn't stay there.

Kailee's Family

He boarded The Rail into the city.

Walked aimlessly for hours.

Then somehow found himself outside Kailee's apartment.

Again.

When Samantha answered the building call system, she immediately heard the pain in his voice.

"I'll get her," she said softly.

The wait felt endless.

Then Kailee appeared downstairs.

The moment Styler saw her, he broke. "I'm scared."

Kailee's expression changed instantly. "What happened?"

Styler could barely get the words out.

"He died."

"Harold died in his sleep."

Kailee immediately pulled him into her arms.

"Come inside."

Upstairs, Samantha, Stephen, and Constance all stopped what they were doing when they saw him.

No one asked unnecessary questions. They simply surrounded him.

Held him.

Fed him.

Stayed near him.

Dinner that night was quiet.

Good food.

Terrible timing.

Rain battered the city windows while soft music played somewhere faintly in the apartment.

The world outside felt hollow.

Alone.

The next morning Styler returned to Harold's house.

It already looked abandoned.

Loose caution tape drifted in the wind outside.

Inside—

The house felt wrong without him.

Too empty.

Too silent.

The following day Styler woke up instinctively shouting: "Harold!"

Silence answered.

Only silence now.

He wandered the city again unable to stay home.

The streets themselves seemed emptier somehow.

The city no longer felt alive.

Wind tunneled between skyscrapers sounding almost like breathing.

Again he found himself at Kailee's apartment.

This time, Samantha sat him down immediately. "We've talked about this," she said gently.

"We don't want you alone in that house." "The thought of it makes me sick." Stephen visibly hesitated.

But eventually sighed.

"Fine."

Samantha smiled softly.

"You stay here from now on." Styler looked overwhelmed.

He didn't know what to say.

So he simply nodded.

The New School Year

The next school year began quickly.

Students returned energized.

Excited.

But Styler felt disconnected from all of it.

He now commuted from Kailee's apartment each day and returned exhausted every night.

Everything reminded him of Harold.

A new teacher introduced herself warmly to the class.

"A new year opens where your last one ended. Congratulations on making it here."

Most students seemed enthusiastic.

Styler barely stayed awake.

Days blurred together.

Rail rides.

Classes.

Assignments.

Silence.

Then one day the teacher announced an upcoming academic break.

"Can't wait to see everyone back afterward!" Students erupted excitedly.

Then suddenly—

"I'm not coming back." The room went silent.

The teacher looked toward Styler carefully.

He met her eyes directly.

"I want to take the equivalency exams." The teacher paused.

"I know what you've been through this year." Then softly:

"If you're sure… we'll arrange it." Styler's Fifteenth Birthday.

Back at Kailee's apartment, Samantha went all out preparing for Styler's fifteenth birthday.

Blue and dark green decorations covered the apartment.

Not ordinary blue.

A vivid sky blue.

The kind Harold once said reminded him of possibility.

The next morning everyone shouted together:

"Happy Birthday!"

Styler smiled weakly.

But inside, he only wanted one thing.

I just want Harold back.

At breakfast he finally admitted it aloud.

"He was my grandfather…" Then quietly:

"My father too."

"My best friend."

Kailee squeezed his hand.

"Souls live on," she whispered.

"They're never completely gone."

Later that evening Kailee took him to a newly popular restaurant operated almost entirely by cyborgs and androids.

Autonomous chefs prepared grilled meals behind transparent cooking stations.

The plates themselves were designed with indented placements for utensils and serving portions.

Everything felt futuristic.

Efficient.

Perfectly organized.

But Styler barely noticed any of it.

He stared down at the table while speaking.

"I'm not going back to school." Kailee frowned immediately. "You have to."

"Why?"

"What are you going to do?" Styler finally looked up.

"First… I'm taking my equivalencies." Then after a pause:

"I've been thinking."

"Either robotic repair work for the military…" "…or university as an early entrant."

Kailee studied him quietly.

For the first time since Harold's death—Styler sounded certain about something again.

19

INTEGRATE

Styler finally decided it was time.

If he was truly going to live at Kailee's apartment, he needed to move at least part of his life out of Harold's house.

Even thinking of it felt wrong.

The house sat quiet around him as he sorted through the contents of his bedroom.

Old tools.

Drawings.

Half-finished ideas.

Pieces of projects Harold had once stood beside him building.

Every object felt alive with memory.

There was far more than he could ever take with him.

And somehow that made leaving harder.

A voice came from the doorway.

"Not easy, is it?"

Stephen stood there quietly, hands in his pockets.

His tone was calm, slower than usual.

"The truck's outside," he added. "Though from the look of it, we may not need much space today."

After a pause he continued:

"Anything else you want later; we'll put in storage."

Outside, several autonomous androids detached themselves from the transport truck and entered the house.

They waited silently for instructions.

Point at something.

Name it.

Or simply gesture.

They carried whatever was requested.

Styler wasn't taking much.

One of the androids carefully lifted his nightstand and carried it outside.

Several wheeled bags automatically followed Styler while he carried a box of personal belongings in his arms.

"Leaving Harold's engineering projects behind..." Styler said quietly, "...that's the hardest part."

He looked back toward the workshop.

"It almost would've been easier to take none of it than only some of it."

Then Fire appeared.

Harold's deep red-coated dog walked slowly up beside him.

Styler crouched and scratched behind his ears one last time.

"Goodbye, Fire."

The dog stared at him silently as though understanding more than an animal should.

The androids reattached themselves neatly into the recessed compartments along the truck's sides.

Then the vehicle pulled away toward the city.

The drive back to Kailee's apartment was quiet.

Construction vehicles passed occasionally.

Farsers trimmed roadside growth in the distance.

The world continued functioning normally despite how much had changed for Styler.

A New Home

When they arrived, Styler immediately approached Samantha.

"Thank you... for letting me stay here."

Samantha smiled warmly.

"You're family."

Then casually:

"You'll have the spare room."

Stephen led him down a long white corridor toward the far side of the apartment.

"And this," Stephen said, opening the door, "is your room."

Inside already sat a furnished bed, dresser, shelving, and workspace.

Everything looked clean.

Prepared.

Intentional.

Styler hesitated.

"I want to keep my nightstand." Stephen nodded immediately. "Naturally."

Then his expression softened.

"Styler… don't think of this as losing something." "This is a step forward."

Later that evening, Styler unpacked quietly inside his new room.

The unfamiliar silence startled him when Stephen suddenly spoke behind him.

"If I told you life was easy," Stephen said, "I'd be lying to you."

Styler looked over.

There was an awkward stillness between them.

Not hostile.

Just uncertain.

Then Kailee entered without hearing the conversation.

"Dad," she said lightly, "let me have Styler to myself for a while."

Stephen smirked faintly.

"Of course."

First Night

It was Styler's first night officially living at Kailee's apartment.

Neither of them wanted to sleep.

They joked around like children trying to avoid bedtime.

Styler sat cross-legged across from her grinning.

"I'm excited."

Kailee groaned dramatically.

"Ugh."

"Go to sleep, Styler." "No."

"Yes."

She laughed.

"I've got plans tomorrow. Will you survive here without me?"

Styler leaned back smugly.

"Copesthetic."

Kailee rolled her eyes.

"That's not even a word." Kailee's Friends.

The next morning sunlight flooded through the apartment's twentieth-floor windows.

Kailee spent the morning with her friends Elizabeth and Sara while Styler stayed behind resting.

The moment Kailee arrived, Elizabeth smirked knowingly.

"You've got boy smell written all over you."

Sara laughed.

"So who is he?"

Kailee tried not to smile too much. Sara continued teasing:

"Erase the romance."

"Why are you even with him?" Kailee giggled despite herself. "I don't know…"

Then softly:

"I guess I fell in love with him along the way." The word love surprised even her.

Still, the teasing made her uncomfortable.

She didn't like hearing Styler criticized.

Soon afterward the girls hugged goodbye and went separate ways.

Styler Gets Sick.

Outside, snow drifted through the city.

Inside, Styler sat sweating heavily.

He looked awful.

Pale.

Exhausted.

The holocaster's sensory therapy wasn't helping.

His stomach twisted constantly with nausea.

Kailee decided to make her own remedy.

Using Samantha's Shopper, she pulled fresh vegetables and fruits directly through the screen interface.

She dried them carefully at low temperature, ground them into powder, mixed vitamins into the blend, then created a nutrient drink for him.

Styler drank it immediately.

The next morning he felt dramatically better.

Reenergized.

Clear-headed again.

Pawning, Antiquing, and Inventing.

That Saturday, Styler suddenly woke up excited.

"Today's the day!" Kailee looked amused. "What day?"

"I'm finally showing you how I make money."

They walked through the city boulevard together surrounded by fashion crowds, autonomous vehicles, and glowing storefronts.

Electric cars raced one another down nearby lanes. "We're going to work," Styler announced proudly. "Pawning."

Kailee blinked.

"Pawning?"

"That sounds dangerous." "What if you lose money?" Styler grinned.

"We're not going to lose money."

"After pawning, we're going auctioning." "And antiquing."

Kailee laughed.

"You do not seem like the antiquing type." "Maybe that's why I'm good at it."

Jake's Pawnshop.

Eventually they arrived at Jake's Pawnshop.

The shop remained dimly lit but crowded with customers. The moment Jake saw Styler, his face lit up.

"Long time no see!"

Then his expression softened.

"Your grandfather's passing hit me too."

Kailee.

"He was a good man."

Jake leaned against the counter.

"So what now?" Styler shrugged.

"Thinking about university."

"Or military robotic repair work." Jake snorted.

"University sounds expensive." "Military sounds dangerous." Then firmly:

"If you choose university, I'll pay half your tuition." Styler stared.

"No, no, you don't have to—" "I'm not asking."

Styler smiled awkwardly before gesturing toward

"This is my girlfriend, Kailee." Jake grinned warmly. "Pleasure to meet you."

Then Styler spotted something immediately.

An old yellow-and-green guitar displayed on a pedestal.

"Oh Jake…"

He pointed toward it.

"I want that guitar back." Jake smirked knowingly. "Eighty-two hour credits." Styler didn't hesitate.

"I'll take it."

RE and InventShop.

Next, they visited RE, a store entirely dedicated to only

recycled goods and reclaimed materials.

Styler browsed carefully before selecting a piece of exotic grawn leather harvested from an off-world animal species.

"Perfect," he said happily.

Then came his favorite place:

InventShop.

The massive public workshop hummed with machinery.

Lathes.

Presses.

Drills.

Industrial printers.

Inventors worked alongside one another sharing tools and ideas.

Styler scanned in immediately.

Then showed Kailee how to gain access. "Shout your name into the scanner." "Kailee!"

Access granted.

Using the old guitar, Styler began modifying it into a hybrid instrument.

Part stringed.

Part woodwind.

He carefully sawed openings into the body.

Then drilled adjustable airflow holes.

"Hey Kailee," he said, handing her tools, "drill these holes and attach the leather."

She did surprisingly well.

Styler smiled proudly.

"Nice job."

Nearby, another patron made a dangerous mistake and automated safety cameras instantly shut down every machine in the workshop.

Moments later operations resumed normally.

Auction House.

Afterward they visited the auction house.

Although the auction hadn't started yet, the attendant still allowed them inside.

Styler proudly showed Kailee around like a tour guide.

Then he approached the auctioneer.

"I've got a house to sell." "And a Morrison Mod 5." "And this custom guitar." Soon the auction began.

The house triggered intense bidding immediately.

"305 thousand hour credits!" "Sold!"

Then the Morrison Mod 5 sold for thirty-five thousand five hundred hour credits.

Finally the custom guitar sold for a massive profit as well.

Afterward Styler sat quietly.

"I can't believe how much money I have now." Then more softly:

"Harold would've hated me selling the house." "But I can't maintain it alone."

Fill.

As evening settled, they stopped at Fill — the same place where they first met.

They even sat at the same table.

Styler ordered two twelve-hour Fill's so neither of them would be hungry until morning.

Then finally he said it aloud:

"I've decided."

"I'm going to university." Kailee smiled instantly. "I hope it's nearby."

"My parents are going to want to visit you constantly." Then Styler suddenly stood.

"Wait here."

He disappeared briefly into a nearby store.

When he returned, he looked nervous.

"I have something for you."

From his pocket he revealed two interlocking rings he had forged himself using embedded electronics and a custom casting mold.

The rings locked together seamlessly.

One hidden feature allowed either ring to silently alert the other wearer during emergencies.

Styler carefully separated them.

"This one's yours."

He slid Kailee's ring gently onto her finger. "It's a commitment ring."

Then he placed the matching half onto his own hand.

"And if you ever need me…" He smiled softly.

"Just press the ring."

20

UNI

Morning sunlight poured through the apartment windows, bright and sharp against the skyline.

Styler woke up unusually energized.

Today was important.

He was finally going university shopping.

At breakfast, everyone sat around the table while Styler quietly ate a serving of rehydrates—compact space-ready food people occasionally consumed even on Earth to prepare for emergency launches or extended orbital travel.

Living inside a launch-capable skyscraper made preparedness normal.

Samantha yawned while pouring herself a drink.

"So," she said sleepily, "what's everyone doing today?" Constance answered politely.

"My agenda's full at home."

Stephen adjusted a holo-report beside his plate.

"I've got to file a structural report on the collapsing underground landfill systems."

Styler stayed oddly quiet.

Not even chewing loudly. Kailee noticed immediately.

"What are you doing today, Styler?" Styler swallowed.

"Well…"

A grin broke across his face.

"I'm checking out my top two university choices."

Then quickly:

"But I kind of want to do it alone." Stephen nodded approvingly.

"Fair enough."

"If you need help or advice, ask." Kailee tried sounding casual.

"I'm probably just staying here today." Then added in a slightly offended tone: "Maybe going out with friends."

As if silently asking why she hadn't been invited.

Styler awkwardly avoided eye contact and stood up. "See everyone later."

Choosing a Future.

The city streets were alive with color and movement as Styler walked alone toward one of the universities he'd been considering.

Tall buildings reflected sunlight in every direction. Pedestrians crossed overhead walkways.

Transit rails thundered past between towers.

The city looked successful.

But the closer he got to the downtown university district, the more something felt off.

A man standing outside a storefront suddenly called toward him.

"Don't go to that degree mill." Styler slowed.

The man pointed toward the nearby city university
towers.

"Extortion scheme."

"Look around. You see any substance for miles?" "Just buildings and food."

"Everybody already has those."

He stepped closer.

"You want a real future?" "Go to Stout University."

"There's a quarry nearby. Morrison's manufacturing plants. Machine shops. Six major headquarters within commuting distance."

"Think of the possibilities." Styler smiled immediately.

"That was already one of my top choices."

As he continued through the city, he noticed more homeless people than he expected.

People sleeping beneath glowing advertisements.

People ignored by crowds.

The city suddenly felt artificial.

Like appearance without foundation.

He stopped walking.

Then quietly turned around toward the rail station.

Decision made.

The Train to Stout.

The train accelerated beyond five hundred fifty miles per hour through endless forests and open land.

Rain streaked across the ceiling windows while the landscape blurred into green and gray.

Styler watched everything carefully.

Wide untouched land stretched for hundreds of miles.

No suburbs.

No crowding.

Just possibility.

He saw exposed rock formations in the distance and immediately thought of engineering materials.

Resources. Infrastructure.

Opportunity.

"Next stop: Stout Station." Moments later:

"All passengers departing at Stout Station please disembark."

Styler jumped up instantly.

Outside the station, autonomous university taxis waited in neat rows displaying Stout's colors—red and blue—across their external holo-screens.

Styler climbed into one.

Within minutes he saw the university.

And immediately understood why people respected it.

The campus looked massive yet compact.

Multiple buildings interconnected by elevated catwalks.

Two skyscrapers curved together in an oblate circular formation.

Underground entrances.

Engineering workshops. Agricultural sectors.

Research towers.

Everything designed with purpose.

Even the architecture looked intelligent.

Styler entered through a large welcoming building featuring heavy wooden double doors etched with geometric patterns.

Inside sat rows of seating illuminated by shifting colored light.

At the center waited a robotic intake system. "Take a seat," the intake unit announced. "You are number thirty-eight."

"Proceed through the right-side doors when called." Styler sat beside another boy roughly his age.

"What number are you?"

"Thirty-seven," the boy answered.

Styler laughed.

"I'm thirty-eight."

"What are you studying?" Styler asked. "Hopefully engineering."

Styler immediately grinned.

"Me too."

"Thirty-Seven. First door." The boy stood.

Before leaving, Styler quickly asked: "What's your name?"

"Forester."

Then he disappeared through the first door. Moments later the robotic voice blared dramatically: "Thirtyyyyy Eiiighttttt. Third door."

Negotiating Tuition.

Inside sat a large engraved wooden desk.

Behind it floated nested chairs stacked elegantly into one another.

An exhausted academic advisor gestured Styler inside.

"I want engineering," Styler said immediately.

The advisor nodded.

"Tuition is forty-two thousand hour credits annually." "Room and board roughly fifteen thousand unless

commuting."

Styler smiled.

"How about half off?" The advisor blinked.

Styler leaned forward.

"People negotiate cars." "Why not tuition?"

"Why not careers?"

The advisor looked stunned. Styler continued confidently:

"If nobody in business classes negotiates tuition prices, then honestly, your business program failed."

The advisor actually started laughing.

Then he tried negotiating back.

"What if I raise your class costs instead?" Styler shrugged casually.

"Then I'd negotiate harder."

Eventually the advisor disappeared into a side room to speak with his supervisor.

Styler waited quietly while examining something strange sitting beside the desk.

An organic computer.

Part biological mass.

Part holo-interface.

It pulsed subtly as though alive.

Processing.

Watching.

Learning.

Styler couldn't stop staring at it.

For a moment he genuinely felt like it was reading his thoughts.

The advisor returned alongside his supervisor.

Without warning, the supervisor handed Styler a red-and-blue Stout sweatshirt.

"Deal," the supervisor said.

Styler smiled knowingly.

Hard bargain won.

The advisor sighed dramatically.

"Are you going to negotiate every semester?"

Styler pulled the sweatshirt over his head proudly.

"If you knew me," he replied, "you wouldn't be surprised."

Kailee at Home

Meanwhile, back in the city, Kailee spent the afternoon playing Teletag with Constance.

Using their holocasters, both girls teleported short distances around the apartment complex grounds while trying to tag one another.

Kailee laughed while teleporting across a courtyard.

"I'm after you!" Constance sprinted away.

Kailee teleported again — barely missing her.

Then finally jumped directly beside her and tapped her shoulder.

"Got you!"

"You're it!"

Constance groaned loudly.

Then Samantha shouted from the apartment windows: "Inside!"

Constance protested immediately.

"Why do we always stop when it's my turn?!"

At that exact moment Kailee noticed Styler approaching wearing his new university sweatshirt.

"STYLER!"

He spread his arms proudly.

"I'm enrolled."

Kailee laughed.

"I can't wait until I go to university." Then dramatically:

"And I'm going to the best university in the universe." Styler pointed at his sweatshirt.

"Excuse me. Stout is the best." Kailee smirked.

"On this planet."

One Last Date

Within a week the semester would begin.

Styler spent days packing and preparing.

Kailee hated how quickly time was moving.

Before long she finally said:

"Before you leave…"

"I want us to do something fun together." "A real memory."

Styler hesitated.

"I've got so much to do." Kailee crossed her arms. "Fine."

"Then I'll just support you quietly." That guilt worked instantly.

Styler laughed.

"Okay, okay."

"One more date." Medieval Travel

Together they packed snacks—cheeses and preserved meats from Torquins—and headed toward Time Travel Avenue.

After passing several venues they finally entered Medieval Travel.

The building looked intentionally ancient.

Tapestries covered the walls.

Reclaimed wood beams crossed the ceiling.

Decorative swords and crossbows lined the waiting area.

The attendant smiled warmly.

"You two look like experienced travelers." "Interested in the thirteen hundreds?" Kailee answered immediately.

"That's exactly why we're here."

Soon both changed into rough period-appropriate clothing.

Animal skins. Dirty fabric.

Heavy wear.

They both looked horrified.

"We paid for this?" Kailee asked disgustedly.

Finally the attendant gestured dramatically.

"Ready?"

"Ready!"

The orange waypoint activated.

And the world folded around them.

The Bow Brothers.

They arrived somewhere in the 1200s.

Walking along a muddy trail ahead stood a pale blue-eyed man practicing with a strange split recurve bow.

Styler immediately noticed something unusual.

"Two arrows. One hold." "Clever."

The man grinned.

"Bet you haven't seen this."

Then suddenly split the bow itself into two sword-like blades and pointed them toward Styler and Kailee.

"So what do you want?"

Another voice called from the woods.

"Need help there, brother?"

A second man emerged carrying an even more elaborate bow capable of firing three arrows simultaneously.

Together they were known as the Bow Brothers.

Ericc and Aaron.

The older brother eventually relaxed.

"Feeling generous today." "You two want a drink?"

War in the Thirteenth Century

The brothers led Styler and Kailee through forests toward a walled village.

On the way they encountered robbers.

The Bow Brothers handled them brutally and efficiently.

One lost an ear before the fight even properly began.

Styler realized immediately these men survived through violence.

At the village inn, the group drank together while learning the town expected an incoming winter attack from neighboring starving kingdoms.

A call to arms had already begun.

Within days Styler and Kailee trained alongside the villagers.

Bows.

Axes.

Spears.

Sword fighting.

Mud covered everything.

The Bow Brothers even gifted them weapons.

Eventually war came.

Hundreds of attackers descended upon the village supported by a catapult.

Flaming projectiles crashed into buildings.

Men screamed through smoke.

Styler and Kailee fought beside Ericc and Aaron.

Kailee shot one attacker directly through the eye. Styler helped destroy the catapult using flaming arrows.

The Bow Brothers moved through battle like legends.

Splitting bows into blades.

Killing attackers in motion.

Protecting the town.

Eventually the attackers retreated.

The village survived.

Barely.

That night the four sat exhausted outside the damaged inn.

Aaron drew a circle into the dirt using one of his split bow-swords.

Ericc finally spoke.

"We should call ourselves something." "What are we?"

Kailee looked toward Styler. "What makes our culture special?" Ericc answered first.

"We invent." Aaron nodded. "We adapt."

Styler quietly smiled.

Together they placed hands inward.

"We are Arlenish."

Kailee looked toward Styler in disbelief.

"We were already Arlenish in the twenty-second century." Styler whispered back:

"Our relatives must've carried it forward."

Eventually both felt the familiar pull of time drawing them home.

They located an orange waypoint and stepped through together.

The attendant greeted them casually upon return.

"About time I pulled you two back." Kailee laughed breathlessly.

"What a trip." University Begins.

Soon afterward Styler boarded the train toward Stout for the start of the semester.

The train overflowed with students and autonomous luggage.

Excitement filled the air.

At campus his dorm room overlooked endless wilderness beyond the university outskirts.

Untouched land stretched beyond civilization.

A blank canvas.

Potential.

Later that evening he met his roommates.

Forester—the same boy from admissions.

And Dean—taller, darker-featured, already looking academically prepared.

All three studied engineering.

That night the dining hall served hybridized plant-animal

food engineered for efficiency.

Meanwhile far away in the city, Kailee attended a concert by Repay! with her friends.

Afterward she overheard the lead singer paying reporters and photographers for publicity.

The illusion of fame cracked instantly.

"People don't just make you famous," the singer admitted casually.

"They make themselves famous first."

Kailee quietly walked away realizing even celebrity was manufactured.

Back at Stout, Styler finally prepared for his first university class.

Social and Introductory Course 0101.

Inside the circular lecture room only a handful of students waited.

An older blonde-haired professor entered.

Blue-eyed.

Still working into her seventies.

And the moment she stepped into the room, the entire atmosphere changed.

Lunch ended too quickly for Styler's liking

The engineering shops were becoming the only places on campus where he truly felt awake.

The only places where thought turned into something tangible.

Something useful.

Back inside Supplemental Engineering Shop Course 3113, machinery roared back to life.

Hydraulic presses slammed rhythmically. Cutting lasers hissed through metal.

The air smelled faintly of hot steel, lubricants, and ozone.

Ryan, Forcster, and Parker were already gathered around Styler's workstation when he returned.

Parker pointed at the wire spool.

"You really think this thing could change something?" Styler immediately answered:

"Everything uses wire."

"Transportation. Robotics. Housing. Spacecraft. Medical equipment."

"If something that common becomes even slightly better, then everything becomes better."

Ryan folded his arms. "And cheaper." Styler grinned.

"And cheaper."

Building Something Real.

Styler carefully unrolled another section of prototype wire across the table.

Inside the transparent casing, the liquid-metal core shifted almost organically beneath the surface.

Forester leaned closer.

"That still looks strange."

"It's supposed to," Styler replied.

"If it looked normal, it probably wouldn't matter." Ryan started sketching rapidly across a drafting sheet. "What if the production line rotated the casing while the liquid core was injected?" Parker nodded.

"That'd prevent uneven distribution." Styler's eyes widened.

"That's actually genius." Ryan smiled proudly.

"See? I invent too."

For the next hour the four worked almost nonstop.

They forgot about the clock entirely.

Styler adjusted conductivity tests.

Parker refined containment thickness.

Ryan redesigned reel-feed automation.

Forester unexpectedly solved stabilization issues by suggesting segmented pressure chambers.

The project was no longer Styler's alone.

It was becoming theirs.

And somehow that made it better.

Across the room the Shop Professor silently watched them.

Not interfering.

Just observing.

The way an older craftsman watches younger ones discover purpose for the first time.

Eventually the professor approached their workstation.

"You four look dangerous." Styler laughed.

"Dangerous successful or dangerous destructive?" The professor smirked.

"Usually both."

He picked up the wire carefully. Bent it.

Twisted it.

Tested flexibility.

Then looked genuinely impressed.

"You know what most students do in this room?" He gestured broadly.

"They make useless projects."

"Fancy shapes. Decorative nonsense. Things nobody needs."

Then he held up the wire.

"This solves an actual problem."

Ryan immediately puffed up with pride.

Forester tried hiding his smile.

Parker looked stunned.

Styler stayed strangely quiet.

Because suddenly the project felt real.

Too real.

Fear of Success.

The Shop Professor lowered his voice slightly. "You understand what happens if this works?" Styler nodded cautiously.

"We make hour credits?"

The professor shook his head.

"No."

"You become visible."

The room quieted around them.

The professor continued:

"If your invention gets adopted…" "Factories use it."

"Military uses it."

"Space infrastructure uses it…"

"Then people start depending on you." Styler's expression slowly changed.

The professor noticed immediately.

"You can't become important without becoming a target too."

For the first time all day, Styler looked uncertain.

Not because he doubted the invention.

But because he suddenly understood the weight of creating something meaningful.

Evening on Campus.

By evening the campus glowed beneath red and blue lighting.

Students crossed the elevated walkways between buildings.

Some laughed loudly.

Others hurried toward study halls.

Far above them, fireworks from orientation week still occasionally exploded across the night sky.

Styler walked back toward the dorm carrying pages of diagrams beneath one arm.

His mind wouldn't slow down.

Inside the dorm Dean was already studying.

Forester arrived shortly afterward carrying engineering materials.

Ryan and Parker eventually stopped by too.

Before long, all four were crowded into the room discussing designs.

Dean finally looked up from his coursework.

"You people ever stop engineering?" Styler answered instantly.

"Hopefully not."

Ryan pointed toward Styler's drawings.

"If this thing actually works on industrial scale…" "We could start a company."

Forester laughed nervously.

"We haven't even passed the semester yet." Parker leaned back in his chair.

"That's the best time to start." Styler stared at the diagrams quietly.

Company.

Factories.

Production.

Leadership.

Responsibility.

The future suddenly felt very close.

Kailee in the City.

Meanwhile, back in the city, Kailee sat at the apartment window staring out over the skyline.

The apartment felt quieter without Styler around constantly.

Even Samantha noticed it. "You miss him already?" Kailee tried sounding casual.

"He's only been gone like… a day."

Constance immediately shouted from across the room: "You totally miss him."

Kailee threw a pillow at her.

Later that evening Kailee used her holocaster to call Styler.

The connection shimmered alive beside his dorm bed.

Kailee's holographic image appeared smiling. "How's university life, genius?"

Styler dramatically groaned.

"I think I accidentally started a business." Kailee burst out laughing.

"Already?!"

Styler rubbed his eyes. "I made a wire." Kailee blinked.

"A wire?"

"Yes."

"A really good wire."

Kailee smiled warmly at him.

"You sound happy." Styler paused.

Then quietly admitted:

"I think I am."

The Weight of the Future.

After the call ended, Styler sat alone by the dorm window.

Outside stretched untouched land illuminated beneath moonlight.

Open wilderness.

Future cities.

Future factories.

Future homes.

Potential.

He thought about Harold.

About the Underground.

About the Absent.

About survival.

About invention.

About responsibility.

Everything felt connected somehow.

Then Styler quietly whispered to himself: "To stand the test of time…"

"…you first have to stand for something."

And for the first time since arriving at university, he finally began understanding what Harold had been trying to teach him all along.

Parker leaned back against the shop table and said excitedly,

"Hey, I got an idea for a new propulsion system utilizing side flanges that open to get non-linear thrust."

The Shop Professor immediately perked up.

"Now that," he said while pointing, "is exactly what I want to hear."

He paced slowly between the machinery while talking to the class.

"Everyone has ideas. Most people can even write them down, which is already better than average. But ideas alone don't matter."

He slapped the side of a machine for emphasis.

"If people can't touch it, test it, use it, break it, buy it, or mass produce it — then it's still just a thought."

The class quieted.

"At least you students aren't becoming those paranoid types that hide every idea they have because they're terrified someone might steal it."

He laughed harshly.

"What good is an invention nobody ever sees?" The professor

pointed around the room.

"You think factories are built by one genius hiding in a basement?"

"No."

"You have to share your ideas with thousands of people before a hundred even care enough to help you."

"And only a handful of those will stay long enough to build something real."

Styler absorbed every word.

Because deep down he already knew the professor was

right.

Huddle Time.

Styler clapped his hands together.

"Alright team. Huddle time."

Ryan, Parker, and Forester gathered closer around the workstation while machinery thundered in the background.

Styler carefully held up the wire spool.

The liquid-metal core shimmered faintly beneath the transparent casing.

"Look guys," Styler began, "I think we can actually pull together a business here."

He rotated the wire slowly.

"What I made is simple."

"That's the best part."

"It's easy to manufacture. It's cheaper. It's more conductive. And every single industry on the planet uses wire."

Parker nodded slowly. "Transportation." Ryan added:

"Military."

Forester:

"Space infrastructure."

Styler smiled.

"Exactly."

The Reality of Risk.

Parker crossed his arms thoughtfully.

"My propulsion idea's probably too expensive right now anyway."

He pointed toward Styler's prototype. "But this? This actually has a chance."

Then his expression shifted more serious.

"If we're really doing this…" "…we're going to need luck."

Ryan leaned forward in his chair.

"No."

"We're going to need sacrifice." The room fell quieter.

Ryan looked directly at Styler.

"You willing to risk your house on this?" Then more bluntly:

"If this fails… are you going to let me live with you?" Styler laughed at first.

But Ryan didn't.

Ryan continued:

"This isn't real estate."

"We're not flipping buildings." "We'd be building production." "Factories."

"Supply chains." "Employees."

"What if we lose everything?"

"What if we don't even end up with homes ourselves?" Forester spoke calmly.

"If we do this…"

"…it has to stay honest." Ryan finally smirked.

"You really want to be the entrepreneur taking all the blame?"

He shrugged. "Alright then." "Let's do it." The Proposal.

Later that class, Styler approached the Shop Professor carrying a thick stack of diagrams.

Each page detailed measurements, automation sequences, reel systems, material flow, and production methods.

The professor's eyebrows rose. "You actually finished all this?" Styler nodded proudly.

"And they helped."

He gestured toward Ryan, Parker, and Forester. The professor flipped through pages rapidly.

His expression gradually shifted from curiosity…

…to genuine interest.

"There's a manufacturing facility partnered with the university a few miles away," the professor finally said.

"They may actually be interested in this." Ryan nearly shouted:

"For real?"

The professor nodded. "We'll go next class." Styler's heart skipped. The professor continued:

"It's mostly automated production, but people still oversee operations."

He tapped the diagrams.

"If your process works with their machinery…" "…this could move very quickly."

The Night Before.

That evening the dorm room was pure chaos. Parker paced in circles.

"We're gonna do this."

"We are actually gonna do this." Ryan practiced introductions aloud.

Forester kept reorganizing diagrams nervously.

Styler sat on the edge of the bed staring at the wire prototype.

Almost afraid to blink.

Ryan pointed at everyone.

"Okay listen."

"Factory workers matter."

"Foremen matter."

"Engineers matter."

"Don't act arrogant."

"Be sociable."

"Connections matter more than intelligence sometimes."

Parker jokingly raised his hand.

"What if I accidentally become a billionaire tomorrow?"

Ryan threw a pillow at him.

"Then buy me a house."

Eventually they broke apart for the night.

But Styler barely slept.

His mind stayed trapped in tomorrow.

Social and Introductory Course Number 0101.

The next morning felt painfully slow.

Styler sat half-awake while the professor struggled with her holocaster interface.

Styler leaned toward another student and whispered: "If I'm better with technology than my teacher…" "…I should be getting paid tuition instead."

The student laughed quietly.

Styler continued under his breath:

"The networking and ideas I bring to these classrooms should count as labor."

Before anyone could respond, the professor clapped her hands.

"Everyone settle down."

"Today we have a guest speaker."

A career-transition specialist entered the room.

An older woman with a calm but tired look in her eyes.

She introduced herself simply:

"I'm an engineer turned doctor." The room stirred.

"One person can become many things," she explained. "Your major does not define the totality of your life." "People evolve."

"Industries evolve."

"Sometimes your first dream isn't your final one."

Styler listened for awhile…

…but slowly his attention drifted.

His thoughts kept returning to the factory trip.

To production.

To possibility.

To whether his invention might actually become real.

Factory Visit – Engineering Shop Course Number 3112.

Later that day the entire class boarded an automated transit bus beneath campus.

The ride only lasted a few minutes.

But for Styler it felt monumental.

The Shop Professor casually spoke during the ride: "Administration tried stopping this trip."

The class groaned.

The professor smirked. "But I ignored them." A few students laughed.

"I wasn't letting you guys miss this." The factory itself was enormous.

Built from reclaimed industrial materials.

Solar-paneled walls.

Recycled steel framing.

Rail access stretching behind the property.

Styler immediately noticed logistics before anything else.

Rail systems.

Truck loading bays.

Material storage zones.

Efficiency.

Inside, robotic assembly systems moved continuously.

Hydraulic arms welded components.

Conveyor systems glided endlessly.

Human workers supervised the larger processes.

One worker shouted jokingly from across the factory: "I have no clue what I'm making!"

"But I'm making it anyway!" The class laughed.

Another worker nearby muttered to a coworker:

"I should've used all this ambition somewhere besides school."

Styler overheard every word.

The Meeting.

The Foreman greeted the class personally.

The Shop Professor quickly pulled Styler forward.

"This," the professor said proudly, "is Styler."

"He developed a wire design that could outperform current market standards."

The Foreman looked interested immediately.

Styler handed him both the schematics and a prototype sample.

"I also designed the automation method."

The Foreman turned the wire over carefully.

Then glanced through the diagrams.

Ryan nearly got flattened by falling debris from above. "What kind of safety standards are these?!" he shouted.

The factory workers burst out laughing.

Styler ignored the distraction entirely.

"You already have railway access to raw materials," he explained to the Foreman.

"You've got automated fabrication."

"You'd only need the custom reel and motor systems integrated into the existing line."

The Shop Professor grinned proudly.

"Nothing gets by Styler."

The Foreman studied the documents several more minutes. Finally he said:

"Yes."

The entire group froze.

"Yes," he repeated.

"We could manufacture this."

"And if performance tests hold up…" "…this could be worth billions." Parker nearly stopped breathing.

Ryan whispered:

"Oh my God."

The Shop Professor and Foreman stepped aside privately. The group immediately huddled together.

"What are they saying?"

"Do you think they hated it?" "What if they steal it?"

"What if—"

The professor returned before they spiraled further.

"He's willing to dedicate a section of the factory for prototype production."

The room exploded with excitement.

"And," the professor added proudly, "he's allowing us back to inspect operations ourselves."

Academic Advisement.

Near the semester's end, Styler visited the Academic Advisor again.

This time ignoring the queueing robot completely. The advisor looked up immediately and smiled.

"If I knew then what I know now…"

"…life would've been very different." Styler grinned.

"Just saying hi."

Then he carefully placed the wire prototype on the desk.

"Look what I made."

The advisor examined it curiously.

Styler said quietly:

"It's going to change the world."

Then after a pause:

"And I just wanted to thank you."

The Academic Advisor looked genuinely touched.

"For what?"

Styler smiled softly.

"For not treating me like I was crazy."

21
GOING BACK

Styler boarded the train alongside a flood of other university students.

Suitcases rolled autonomously behind their owners, aligning themselves neatly beside seats or overhead compartments. Styler dropped into an open seat by the window while his own suitcase positioned itself automatically nearby.

He slipped on his holocaster and resumed listening to an audiobook—a techno-novel filled with philosophy, invention, and futuristic ideals.

Through the device he heard:

"The key to life is the love we have in ourselves. The love we have for others. How we share our love with the time we have."

Styler quietly stared out the train window while the city approached.

The words stayed with him.

Back in the City

Once the train reached the depot, Styler made his way through the bustling city toward Kailee's apartment.

The moment Kailee spotted him, she shouted excitedly:

"STYLERRR!"

She nearly tackled him hugging him.

Samantha leaned from the kitchen and asked warmly: "Are you hungry?"

Dinner was already being prepared.

That evening the apartment was lively.

Styler and Kailee joked around endlessly at the table.

At one point Styler shoved carrot sticks into his nose and immediately regretted it when he struggled getting them back out.

Kailee laughed so hard she nearly cried. Samantha shook her head.

"You two are hopeless." The Eeeboard.

Later that night Styler sat at the eeeboard in the living room.

The instrument floated slightly above its stand, its illuminated keys suspended in air like holographic glass.

It was the evolutionary successor to the piano and keyboard.

Minimal waste.

Near seamless design.

Able to emulate almost any instrument — and even emotional frequencies themselves.

The eeeboard synchronized directly with holocasters, allowing emotional resonance to be projected through sound.

Styler began playing softly.

The melody started gentle…

…then unexpectedly melancholic.

Kailee suddenly groaned dramatically: "Stop making me cry!"

She rubbed her eyes.

"I know there are cry-keys on that thing, but you don't have to use them."

Styler laughed quietly while continuing to play.

Kailee hated when technology manipulated her emotions too directly.

Even when it was beautiful.

A Different Look.

Later, while Styler was changing clothes in his room, Kailee called through the door:

"Everyone has outfits they never stop wearing." "You get too comfortable. That's a weakness." "Mismatch something on purpose sometime."

A moment later Styler stepped out wearing something entirely different from his usual blue sweatshirt.

A deep red shirt.

Dark blue pants.

Different.

Sharper.

Older somehow.

Styler awkwardly asked:

"What do you think?"

Kailee immediately pointed at him dramatically.

"Fantastic."

"See? You look good in colors besides blue." Styler folded his arms protectively.

"I'm attached to that sweatshirt." Kailee laughed.

"Clearly."

Styler shrugged.

"Red and blue are school colors."

"Thought I'd try something different." Talking About Harold.

The joking mood faded slightly.

Styler grew quieter.

"There's something I wanted to talk to you about." Kailee immediately listened.

Styler sat down slowly.

"I never understood people who time traveled just to revisit dead relatives."

He paused.

"I thought it was dangerous emotionally." "But now…"

He looked down.

"I understand."

Kailee softened immediately. "You want to see Harold." Styler nodded silently.

Kailee answered carefully:

"I'd probably time travel forever just to be with you." Then after a moment she added:

"What about the holocaster past-feature?" Styler shook his head.

"It isn't real enough." He looked at her directly. "Come with me?"

Kailee smiled gently.

"I'll wait outside for you."

"I don't want to interrupt your time with him." Then jokingly:

"I've talked to my dead grandmother like fifty times already."

Time Travel Avenue.

The city was beautiful that evening.

Warm sunlight reflected off towering buildings.

Cloud advertisements drifted overhead.

Automated transit lines hummed constantly between districts.

Walking beside Kailee toward Time Travel Avenue, Styler suddenly said:

"You ever think about everyone moving around this city?"

"All these people riding transportation, walking streets…"

"Where are they all going?" "And for what?"

Kailee smiled.

"The things we never know are usually the most interesting."

Above them a holographic weather broadcaster appeared across the clouds.

"Ever felt like the weather was just for you?" "Maybe it was today!"

The weather truly was perfect.

Kailee suddenly pointed subtly toward a stranger nearby.

"See that guy's aura?" "It's interesting."

Styler nodded thoughtfully.

"Auras are metabolic usually."

"If people consume too much sugar, they sort of…highlight."

"Alcohol makes auras wobble."

"Some people can isolate them easier than others." "In crowded places it becomes sensory overload." He glanced around the city.

"It's like smell."

"Sometimes you enter a packed venue where there should be hundreds of scents…"

"…and you smell nothing." Kailee looked impressed.

"I never thought about it that way."

Visiting Harold.

They arrived at Time Travel Avenue.

Kailee waited outside while Styler entered Time Travel For You.

Moments later he traveled back to the year 2130.

Outside Harold's home, Fire—Harold's genetically altered red-haired dog—bounded toward him happily.

Styler crouched beside the dog, playfully trying to teach him mathematics using hand gestures and nearby objects.

Then he hurried inside.

"It's dusty in here."

Harold immediately replied jokingly:

"You know how much dust I'll make when I die?" Styler winced hard at the comment.

Because Harold already was dead.

Just not yet here.

Styler finally admitted quietly:

"I'm not from this time." Harold stared at him.

Then nodded slowly.

"So you time traveled to see me." Another pause.

"I must be dead then."

Harold's eyes began watering immediately. "I failed you too."

He was thinking about Styler's parents.

His daughter. Everything lost.

Styler interrupted quickly:

"I'm not here to talk about that."

"I wanted to tell you what I've done." "At university."

Harold looked at him carefully. Styler smiled faintly.

"I invented something."

"A factory's picking it up already." Harold listened silently.

Outside with Kailee.

Meanwhile outside, Kailee waited between travel venues.

Nearby a loud voice carried through the plaza.

It was inventor John Williamson speaking publicly to a crowd.

"Don't fear science and technology!" "It's the ultimate celebrity!"

"The first celebrities on television were nerds!" "Because nerds made the televisions!"

Kailee immediately recognized him.

She approached.

"My boyfriend contacted you last year asking about becoming your apprentice." "His name's Styler."

"Why didn't you ever contact him?" John Williamson answered dryly:

"Trace the history of inventions and how they spread." "Then you'll understand why."

A beat later he smirked.

"I was busy."

Kailee groaned loudly.

"Ughhhh."

Nearby another technology demonstration was occurring.

A newer holocaster series was being shown publicly.

The devices reacted almost before users consciously gave commands.

Think of a function—it activated.

Think of a product—it could be autonomously manufactured or 4D printed nearby.

Convenient.

Dangerously convenient.

A man demonstrating the devices proudly declared: "His story is the past."

"Mine is the future." Harold's Advice.

Back inside Harold's home, Styler continued talking excitedly.

He described giant cruise ships.

Self-sustaining greenhouses.

University life.

Engineering.

Factories.

The future.

Then quietly he said:

"Come visit me."

Harold immediately shook his head. "You know we can't do that." Styler insisted:

"You'd be proud of me."

Harold suddenly became serious.

"You're doing this for the wrong reasons." Styler froze.

Harold continued:

"You're trying to impress someone who isn't there anymore."

"And I'm not there."

The words hurt badly. Then Harold added softly:

"Live in the past…"

"…and you become history." Styler stood there silently.

Feeling like he was staring at himself decades older.

Finally Styler whispered:

"You're so much like me."

Or maybe…

"I'm like you."

Then his thoughts spiraled.

Take me back.

Take me back.

Take me back.

Styler abruptly left.

Out the door.

Out the venue.

Back through time.

Kailee Gets Sick.

When Styler returned, Kailee looked terrible.

Pale.

Weak.

Sweating lightly.

Barely standing.

Styler immediately noticed.

"You must've caught something at that concert." Kailee coughed harshly.

"Maybe."

Styler grew alarmed quickly.

"You okay?"

"Not one hundred percent."

She nearly collapsed.

Styler removed his jacket immediately and wrapped it around her shoulders.

She was freezing despite sweating.

"MedBot! My location!" "Scan female, Kailee!"

A MedBot 406 descended rapidly overhead and scanned her body.

Moments later an alert appeared: UNKNOWN ERROR

Styler frowned hard.

"Again!"

The same result appeared.

UNKNOWN ERROR

People nearby started paying attention.

One man nearby—a Speaker of Nadjin—stepped closer.

He studied Kailee carefully. Then said quietly:

"I've seen this before." "It's not good."

Time Sickness.

The Speaker spoke calmly.

"This is time sickness."

"People have been spreading it across time periods." "It worsens."

"You need Plutoxon."

Then in Nadjin he whispered:

"Lw en, ra nesse."

"In time is your answer." Kailee translated weakly:

"He says the future has the cure."

The Speaker continued speaking in Nadjin.

"Za luot heenex, me hon sen gone don, sen so mreon je."

Kailee translated between coughs:

"Equal parts." "Plutox."

"Licem grain." "Kerns fruit."

"Mixed and consumed." Styler asked urgently:

"How do we get Plutox?" The Speaker answered:

"It only arrived once." "On a distant meteor."

"You must travel to another planet." Kailee could barely remain conscious now.

She whispered weakly:

"In the future…"

"…find the right Speaker." "They'll know the stars." Styler looked terrified.

Everything suddenly became time-sensitive again.

Life-or-death sensitive.

He carefully lifted Kailee into his arms.

Her limbs dangled weakly.

Styler held her tightly.

"I lost Harold." His voice cracked.

"I'm not losing you."
He looked back toward Time Travel Avenue.
"I wasn't going back in."
"But my situation changed."
"I'm going to the distant future."
"Not because I want to…" "…because I have to."

22
EVACUATE

Styler stood near the helm of the New Pioneers' ship, gripping the edge of the console while stars streaked past outside the viewport.

Every second felt stolen from Kailee.

Every second felt dangerous.

The flexible screen-panels along the ship's walls shifted colors continuously—sometimes showing advertisements, sometimes camouflage, sometimes star maps layered across the interior itself.

Planner sat calmly in the pilot's position while Reeler lounged sideways in one of the spring-loaded seats with his boots kicked up against the wall.

Welder finally stumbled back into the helm after removing his old blue suit.

Welder shouted dramatically:

"We are alive!"

Then immediately pointed accusingly at the damaged hull monitor.

"Barely."

The New Pioneers. Styler looked exhausted.

Reeler noticed immediately.

"You keep staring at space like it insulted you." Styler answered quietly:

"She's dying."

The joking energy in the room softened.

Even Welder stopped complaining for a moment.

Planner finally asked:

"You really love this girl?"

Styler answered without hesitation.

"Yes."

Welder smirked slightly. "That's unfortunate." Styler frowned.

Welder continued:

"Because love makes intelligent people do incredibly stupid

things."

Reeler interrupted:

"Correction."

"It makes intelligent people do brave things." Planner smiled faintly.

"Same thing sometimes." Approaching the Purity.

Hours later the ship's atmosphere shifted.

Traffic increased dramatically.

Spacecraft of every shape imaginable flooded the surrounding void.

Massive freighters.

Luxury cruisers.

Mining vessels.

Military escorts.

Dark smuggling ships.

Some rotated slowly through space like drifting cities.

Others shot past at terrifying speed.

The New Pioneers had finally reached the outer region surrounding the Purity.

The Purity itself appeared gradually beyond nearby traffic.

Styler stared in disbelief.

It was enormous.

Not quite a planet.

Not quite a station.

The Purity looked like countless cities welded together into one colossal artificial world.

Circular rings rotated around its center.

Docking ports glowed endlessly.

Construction lights flickered across its surface like
stars.

Sections of the station looked decades old. Others looked newly manufactured.

Some areas were beautiful. Others looked lawless.

A place where anyone could disappear. Or be sold.

Reeler whistled.

"Beautiful nightmare."

Planner nodded.

"Every civilization eventually builds one place like this."

Welder added:

"And every civilization regrets it." Kailee on the Dark Ship.

Meanwhile aboard the dark-colored spacecraft, Kailee remained partially paralyzed from sedation. Many spacecraft glowed in the darkness of space from their glow in the dark paint, but not the dark craft Kailee was on.

Her body barely responded.

But she could hear everything.

The kidnappers stood around nearby discussing her openly like cargo.

One woman stared at Kailee carefully.

"She's valuable."

Another man laughed.

"She's more than valuable."

"She's from another time period."

"That alone makes her worth planets." Kailee weakly tried moving her hand. Nothing.

The spacecraft passed through heavy docking traffic toward the Purity.

Outside the viewport, endless lights illuminated the artificial world.

Kailee's breathing worsened.

The purple rash along her neck spread slightly further.

One kidnapper noticed.

"She getting worse?" Another answered coldly:

"Doesn't matter."

"As long as buyers see her breathing." Kailee's eyes watered silently.

She thought only of Styler.

Docking Procedures

Back aboard the New Pioneers' vessel, warning lights flashed softly.

Planner announced:

"We're entering Purity-controlled space."

Immediately a mechanical voice broadcast throughout nearby traffic.

"ALL INCOMING VESSELS DECLARE CARGO, CREW, AND ORIGIN."

Reeler whispered dramatically: "And here comes the bureaucracy." Welder muttered:

"The most dangerous species in the universe." Planner entered clearance codes calmly.

The station AI responded moments later:

"NEW PIONEERS CLEARED FOR TEMPORARY DOCKING."

"BAY 441."

Styler exhaled.

They were in.

Arrival at the Purity.

The ship glided through one of thousands of illuminated docking channels.

Massive support structures surrounded them.

Cargo rails moved continuously overhead.

Smaller vessels detached and reattached constantly like insects surrounding a hive.

The Purity truly never slept.

The docking bay finally sealed around them.

With a heavy mechanical thud, the New Pioneers docked successfully.

Planner stood.

"Welcome to the center of civilization." Reeler corrected him:

"Center of surviving." Welder added:

"Center of terrible smells too." Searching for Kailee.

Styler was already heading toward the exit. Planner grabbed his shoulder firmly.

"You run blindly in a place like this…" "…you disappear."

Styler stopped.

Planner continued calmly:

"The Purity has laws." "Markets."

"Districts."

"People who sell food."

"People who sell weapons."

"People who sell information." "And unfortunately…"

"…people who sell people."

Styler clenched his fists hard. Planner softened slightly. "We'll help you find her."

The Purity Interior.

The moment Styler stepped into the station itself, he was overwhelmed.

The interior of the Purity looked infinite.

Layered city blocks stretched upward and downward simultaneously.

Suspended walkways connected entire districts.

Artificial skies projected different times of day overhead depending on the zone.

Some sections resembled luxury cities.

Others resembled industrial nightmares.

Languages echoed everywhere.

Human.

Nadjin.

Temouth.

Languages Styler had never heard before.

Species from distant planets walked among crowds.

Robots moved constantly between them carrying cargo.

Flying advertisements projected directly into the air.

"Need oxygen modifications?"

"Visit New Eden District!"

"Planetary transport discounts available!"

"Genetic alteration clinics open twenty-four hours!" Styler felt small.

Very small.

The Information Broker.

Planner eventually guided the group into a darker district beneath the main commercial sectors.

The lighting turned dim blue.

Steam rose from vents beneath the floor. Music echoed faintly through narrow corridors. Reeler casually whispered:

"This is where honesty goes to die."

They entered a cramped shop filled with suspended screens and mechanical parts.

A woman sat behind a counter wearing a silver visor across her eyes.

Planner greeted her.

"Indexer."

She answered immediately:

"Planner."

Her head tilted slightly toward Styler.

"You brought trouble." Reeler grinned.

"He came with it." Planner spoke calmly:

"We're looking for a recently arrived girl." "Human."

"Time displaced."

"Likely intended for ransom or sale." Indexer became serious immediately.

"That narrows it down less than you think."

She activated several floating displays. Images of docking records appeared rapidly. Dark-colored ships.

Cargo manifests. Passenger logs. Smuggling routes.

Then suddenly one result paused. Indexer pointed.

"There."

Styler stepped closer immediately. It was the same dark spacecraft. Indexer continued:

"That ship belongs to a faction operating inside Lower Purity."

"They sell rare biologicals." "Time travelers."

"Genetic anomalies." "Historic artifacts." Welder muttered angrily:

"Scum."

Styler asked urgently:

"Where is she now?" Indexer hesitated.

Then answered quietly:

"Sector Nine."

The room went silent.

Even Reeler stopped joking. Styler noticed immediately. "What?"

Planner finally answered:

"Sector Nine is controlled territory." "No laws."

"No station enforcement." "Only gangs."

Reeler added:

"And worse."

Kailee Wakes.

Meanwhile Kailee slowly regained partial movement.

She was no longer aboard the ship.

She was inside a dim metallic room.

A holding room.

Her wrists were restrained lightly to a medical chair.

Machines monitored her condition.

The purple rash had spread further along her arms.

A woman nearby examined data screens.

"She's deteriorating faster." A man answered:

"Doesn't matter."

"She only needs to survive long enough for bidding." Kailee weakly whispered:

"Styler…"

The woman looked at her strangely.

"You really think someone's coming for you?" Kailee slowly opened her eyes fully.

"Yes."

Entering Sector Nine.

Back in the lower districts, Styler and the New Pioneers prepared themselves.

Planner handed Styler a compact atmospheric sidearm.

Styler hesitated.

"I've never used one like this." Welder answered:

"Today's a fantastic day to learn." Reeler tossed Styler a small devicc.

"What's this?"

"Emergency beacon."

"If you get separated…" "…press it."

Planner looked directly at Styler.

"You stay close to us."

"The Purity eats people who wander." Styler nodded.

But internally he only had one thought.

Find Kailee.

Before time ran out completely.

23
THE NEW PIONEER

Styler carefully crushed the Licem grain with the side of a metallic crate lid while the firefight echoed deeper through the storage bay.

The Purity trembled faintly from distant impacts somewhere aboard the colossal vessel.

Warning sirens pulsed overhead.

Red lights flickered.

Kailee barely moved.

Styler's hands were shaking.

Planner noticed.

"Easy."

"You spill the ingredients, she dies." That sentence alone steadied him.

Styler slowly opened his hoodie pouch and removed the glowing fragments of Plutox they mined from the cratered planet.

Even in the dim storage room the material shimmered unnaturally.

Welder backed away slightly.

"That stuff gives me bad feelings." Reeler whispered dramatically:

"Everything gives you bad feelings." Creating Plutoxon.

With them, the MedBot extended several mechanical instruments outward.

"PREPARING COMPOUND."

Styler added the measured Plutox carefully into the MedBot chamber.

Then the Licem grain.

Then the crushed Kerns fruit.

The machine began processing immediately.

Inside the transparent chamber the ingredients reacted violently.

Silver.

Purple.

Blue.

The substances twisted together almost alive.

The MedBot announced:

"COMPOUND CREATED." "PLUTOXON READY."

Styler immediately grabbed the injector.

Planner stopped him.

"You only get one chance." Styler looked directly at Kailee.

Then administered the cure.

Waiting.

Nothing happened.

Styler's stomach dropped instantly.

Kailee remained motionless.

The room felt unbearably silent despite distant gunfire.

Then suddenly—

Kailee gasped violently.

Her entire body arched upward.

The purple rash spread rapidly for one horrifying moment—

—and then slowly began fading.

The MedBot rescanned her.

"TIME SICKNESS STABILIZING."

"PATIENT CONDITION IMPROVING."

Styler nearly collapsed from relief.

He grabbed Kailee tightly.

"You scared me to death." Kailee weakly smiled.

"Took you long enough." Trouble Aboard the Purity.

The reunion lasted only seconds.

Suddenly the entire ship lurched hard sideways.

The lights flickered.

A distorted announcement echoed through the corridors.

"WARNING."

"UNAUTHORIZED COURSE ALTERATION DETECTED."

"PURITY TRAJECTORY FAILURE."

Planner immediately looked toward Styler.

Styler froze.

Then slowly remembered changing the Purity's navigation

path earlier while experimenting at the helm.

Welder stared.

"You touched the controls?" Styler answered quietly:

"…maybe."

Reeler screamed:

"MAYBE??"

Another violent tremor hit the ship.

Cargo containers crashed from upper racks.

Outside the storage bay viewport, stars shifted unnaturally.

The Purity was moving incorrectly.

Fast.

Far too fast.

Planner rushed to a nearby display terminal.

His expression darkened immediately.

"Oh no."

Styler stepped closer.

"What?"

Planner answered:

"You redirected the Purity toward a gravitational collapse zone."

Welder shouted:

"YOU DROVE A CITY-SPACESHIP INTO A DEAD UNIVERSE?"

Reeler pointed accusingly.

"I officially blame the engineer."

Evacuation Panic.

Across the Purity, panic spread instantly.

Passengers flooded the corridors.

Emergency alarms screamed overhead.

Artificial gravity flickered repeatedly.

Some districts lost power completely.

Android staff attempted crowd control.

Autocops deployed throughout the concourses.

The gigantic spacecraft began rotating unevenly.

Far outside the viewports, space itself looked wrong.

Dark.

Distorted.

Almost folded inward.

Welder looked horrified.

"That's not normal space." Planner answered quietly: "No."

"It's collapsed space."

Kailee Recovers.

Kailee slowly regained strength while sitting upright against the cargo crate.

Her breathing normalized.

The color slowly returned to her face.

Styler stayed beside her the entire time.

Kailee looked at him softly.

"You came for me."

Styler answered immediately:

"I always will."

Reeler loudly interrupted:

"Wonderful." "Romantic."

"We are also all about to die."

Escape Plan.

Planner quickly projected a holographic layout of the Purity.

"We need to reach emergency tertiary craft." Welder objected immediately.

"Everyone else is thinking the same thing." Planner nodded.

"Exactly why we leave before panic reaches the hangars."

Forester looked around nervously.

"This ship is huge…"

"How do we even get out?" Styler suddenly focused.

"The morphing concourse." Everyone looked at him.

Styler continued rapidly:

"The Purity changes internal layouts dynamically." "If I can access the structural controls…"

"…I can reroute corridors." Reeler blinked.

"You can hijack the inside of the ship?" Styler answered:

"I think so." Welder groaned.

"That sentence always leads to catastrophe." Running Through the Purity.

The group sprinted from the storage bay into the increasingly chaotic corridors.

Passengers shoved past them. Some cried.

Others fought.

A few simply froze in fear.

The station-sized spacecraft groaned around them like a wounded animal.

Styler used his holocaster repeatedly against locked doors and routing systems.

Entire hallways shifted shape.

Walls folded.

Walkways extended.

Elevators rerouted.

Forester stared in disbelief.

"You're literally reprogramming the Purity."

Styler answered while running:

"I'm trying not to crash it harder."

The Observation Deck.

They finally emerged briefly into a massive observation corridor.

Everyone stopped for just one moment.

Outside the colossal transparent hull, space itself appeared torn apart.

Stars bent unnaturally.

Entire regions looked swallowed by darkness.

Kailee whispered:

"It's beautiful…"

Planner answered grimly:

"It's death."

Suddenly the Purity shook harder than before.

A nearby section of hull ruptured violently.

People screamed.

Emergency shields activated instantly over the breach.

Welder shouted:

"MOVE!"

The group ran again toward emergency craft sectors while the dying megaship drifted deeper toward the collapsing edge of space itself.

The orange glow of the time portal faded behind them.

For the first time in what felt like forever—there was silence.

Not war.

Not alarms.

Not screaming metal.

Just silence.

And breathing.

Real breathing.

The city of year two one three eight moved around them peacefully.

Rails glided overhead.

Artificial clouds drifted slowly across the sky.

Advertisements shimmered from distant buildings.

Someone somewhere laughed.

Someone argued over hour credits.

Someone carried groceries home.

Normal life continued.

Styler stared at it all almost unable to process it.

After seeing collapsing planets, dead galaxies, wars in space, and the Absent beyond time itself—the ordinary felt sacred. Kailee noticed his expression. She softly asked:

"You okay?"

Styler answered honestly.

"I don't think I ever appreciated home before."

Returning Home.

The three of them walked through the city together.

Forester was overwhelmed.

He kept staring upward at every skyscraper and rail system.

"This place feels ancient now." Styler laughed.

"It basically is."

Forester says while looking around:

"I don't even know what year feels real anymore." Kailee smirks.

"That means you're officially a time traveler." Forester quietly smiled at that.

Back at the Apartment.

Kailee burst through the apartment doors dramatically.

Samantha immediately shouted from the kitchen: "You're late!"

Kailee shouted back:

"I was kidnapped in space!" There was silence.

Then Samantha replied casually:

"Well dinner is still warm."

Styler started laughing uncontrollably.

Kailee joined in.

Even Forester laughed awkwardly though he had absolutely no context for this family dynamic.

Stephen walked into the room holding a cup.

He looked at Forester.

Then Styler.

Then Kailee.

Then said:

"I'm not asking."

A New Normal.

Days slowly began returning to normal.

Or at least as normal as possible.

Styler returned to University.

Though now—

everything felt smaller.

Not unimportant.

Just smaller.

A classroom couldn't compare to seeing universes collide.

Engineering diagrams felt different after repairing spacecraft in deep space.

History class felt ridiculous after literally living through future catastrophes.

Still—Styler kept showing up.

Because Harold was right.

Life keeps moving.

Engineering Expansion.

Styler's wire invention continued exploding across industries.

Factories licensed it.

Spacecraft began integrating it.

Entire relay depots switched over to the technology.

His holocaster notifications became endless.

Licensing agreements.

Manufacturing contracts.

Interplanetary requests.

Ryan once stared at the numbers in disbelief.

Ryan shouted:

"You accidentally became one of the richest people alive!"

Styler answered:

"Doesn't feel very different."

Forester immediately replied:

"That's because you're insane."

The New Pioneers.

Far away in the future—aboard their repaired spacecraft—Reeler was still spending money irresponsibly.

Planner sat at the helm reviewing star maps while muttering about "financial instability."

Welder upgraded half the ship simply because he could now afford better equipment.

Reeler shouted through the ship: "I bought a moon!"

Planner screamed back:

"YOU CAN'T MAINTAIN A MOON!"

Reeler proudly responded: "I'll rent it out."

Welder quietly whispered:

"He actually might."

The New Pioneers kept traveling.

Kept discovering.

Kept surviving.

Harold.

One night Styler sat alone looking through old engineering diagrams Harold helped draw years ago.

The house was quiet.

Kailee eventually sat beside him.

Neither spoke for a while.

Then Styler quietly admitted:

"I still miss him every day." Kailee leaned against him. "You always will."

Styler nodded slowly.

"But I think that's okay now."

Final Semester.

The semesters moved quickly after that.

Styler, Ryan, Parker, and Forester became known around the engineering departments.

Some professors admired them.

Others considered them dangerous.

Mostly because they actually built things.

Actually questioned things.

Actually challenged systems.

Professor Tack once pointed directly at Styler during class and announced:

"Most people adapt to the world."

"Engineers attempt to adapt the world to people."

Then he paused.

"And occasionally create entirely new problems."

The class laughed.

Styler did too. Graduation.

Another bright day. Another ceremony.

Except this time—it truly felt earned.

Styler stood beside Kailee outside the university afterward.

Students celebrated everywhere.

Families cried.

Pictures were taken endlessly.

Forester was hugging everyone despite pretending he hated social interaction.

Ryan was already discussing business expansion.

Parker was pitching propulsion concepts to random graduates who absolutely did not ask.

Kailee looked at Styler and smiled.

"So what now?"

Styler looked out across the city.

The rails.

The towers.

The skies.

The world.

Then beyond it.

To everything still waiting somewhere out there.

He answered calmly:

"Now?" "We build."

Above them—through the artificial clouds—a faint shooting star crossed the daytime sky. Or maybe—it was just another spacecraft leaving the planet behind.

QUESTIONS

Q1: Who are the protagonists and antagonists of the novel?

Q2: List three or more devices or technology (can be infrastructure) from the novel you liked and why? Be descriptive and imaginative as to why.

Q3: Name a device or technology (can be infrastructure) in the novel you wish were real, and why?

Q4: Name a device or technology (can be infrastructure) you came up with on your own that you wish were real, and why?

Q5: How could you turn one of those devices into reality?

Q6: Name a location and time period you wish you could time travel to and why?

Q7: Is the novel set in the right year for the technology and social circumstances?

Q8: Did Styler need school to become successful, could he have been just as successful without school?

Q9: Explain the novels' title and then come up with your alternative title.

Q10: Come up with your own group question here and share it.

AUTHOR'S NOTES

IF YOU WISH, you can skip this portion of the novel and only have enjoyed the story but that's up to you. You made it this far. As this is my debut novel, I felt I might as well write something. I'll start with what Sci-Fi means to me. It is the future, not just on the page, but one that is foreseeable in real life. Why couldn't the Sci-Fi book you read, or movie you watch, be the real future one day? A real prediction of it or at least close enough. Sci-Fi helps us simulate hypothetical situations that we may have in the future and need knowledge for.

Technology in Sci-fi means to me that all the pieces of technology, infrastructure, or other things within the book could actually really be made. As in mass-produced, tangible, and commercially implemented for people to use. Sci-Fi has technology people may actually make and need one day.

Supporting a book or other piece of art can be like crowdfunding and crowd promotion. Sci-fi promotes something real, tangible and can be everyday life. Buying entertainment is like crowdfunding what's inside the piece of work and those involved. I like the shared vision of Sci-Fi the most of any genre. If I were funding or crowdfunding anything, it would be the future, which is in Sci-Fi books, music, movies, TV shows, online games, board games, and video games. There are online lists of "movies that got technology right" and there are even books that have gotten technology right too. Although there is a genre for everyone, to me, Sci-Fi is the most important mass media genre. Other genres can be encompassed into Sci-Fi, for example, who says a Sci-Fi doesn't also have romance in it, adventure, or even a Sci-Fi fantasy like some of my other writings.

Sci-Fi writing and media allow people to make a contribution who might otherwise be ignored. This is some of my contribution to technology. Sci-Fi is not quite academic knowledge but it allows open creativity and artistic expression. Sci-Fi occasionally has tidbits of investing information for national and global stock markets or privately because of the technology inside. I personally never feel bad about buying Sci-Fi because you'll occasionally see something back in the form of a tangible product to the world.

I hope to make it onto one of those "Got it right lists" in some way or form. With this book I contributed more than ten pieces of technology and infrastructure that could become real one day. Coming from a country in debt that suffers dependency I try to do my part in any way possible with intellectual property that could be made real and exportable. Even infrastructure from this could have blue prints and invoices sold as an export. It's possible even one significant technological contribution, even in Sci-Fi novel form, will deserve an award for the novel.

Additionally, I didn't want to create something that was language blind and want pieces of my entertainment to transcend as real things or in media like film. I found Sci-Fi writing to make the most sense when it comes to having intellectual property that I can't feasibly make or having more intellectual property than I could ever make in a lifetime. As I wrote this unemployed and don't have a laboratory, workshop, manpower, extreme wealth or other means to make some of these things possible; so they might as well be in a book for you to enjoy. I'm hoping it will become multiple movies one day either animated or live action with real life prototypes shown, used, tested in the films. For awhile I saw this as a book series, either written only by myself or co-written, but its fate is yet to be determined. It is a young adult series in reading level but suitable for all audiences. This book was written without AI and was not ghost written also not co-written, although AI would be the newest technology and attractive for a Sci-Fi novel that features new technology in it. It seemed unjust because the book used my imagination and I'm not AI, I'm human.

Special thanks to all the Sci-Fi writers and authors that laid a foundation for me. If it weren't for them, there would be no genre. Even more so, a special thank you to the people in science and technology, there would be no science fiction genre without you.

Special thanks to of course, my mother, Norine. She isn't quite a Sci-Fi fan but she helped me edit my first draft. The first book is dedicated to you.

Made in United States
Orlando, FL
25 July 2026

84787562R00141